D0007944

COMPUTER-AIDED ENGINEERING
Heat Transfer and Fluid Flow

ELLIS HORWOOD SERIES IN MECHANICAL ENGINEERING

STRENGTH OF MATERIALS
Vol. 1: Fundamentals – Vol. 2: Applications
J. M. ALEXANDER, University College, Swansea
TECHNOLOGY OF ENGINEERING MANUFACTURE
J. M. ALEXANDER, University College, Swansea, G. W. ROWE, Birmingham University and R. C. BREWER
ELASTIC AND PLASTIC FRACTURE
A. G. ATKINS, University of Reading, and Y. W. MAI, University of Sydney
VIBRATION ANALYSIS AND CONTROL SYSTEM DYNAMICS
C. BEARDS, Imperial College of Science and Technology
STRUCTURAL VIBRATION ANALYSIS
C. BEARDS, Imperial College of Science and Technology
COMPUTER AIDED DESIGN AND MANUFACTURE 2nd Edition
C. B. BESANT, Imperial College of Science and Technology
BASIC LUBRICATION THEORY 3rd Edition
A. CAMERON, Imperial College of Science and Technology
SOUND AND SOURCES OF SOUND
A. P. DOWLING and J. E. FFOWCS-WILLIAMS, University of Cambridge
MECHANICAL FOUNDATIONS OF ENGINEERING SCIENCE
H. G. EDMUNDS, University of Exeter
ADVANCED MECHANICS OF MATERIALS 2nd Edition
Sir HUGH FORD, F.R.S., Imperial College of Science and Technology, and J. M. ALEXANDER, University College of Swansea.
MECHANICAL FOUNDATIONS OF ENGINEERING SCIENCE
H. G. EDMUNDS, Professor of Engineering Science, University of Exeter
ELASTICITY AND PLASTICITY IN ENGINEERING
Sir HUGH FORD, F.R.S. and R. T. FENNER, Imperial College of Science and Technology
DIESEL ENGINEERING PRINCIPLES AND PERFORMANCE
S. D. HADDAD, Associate Professor, Western Michigan University, USA, and Director of HTCS Co., UK, and N. WATSON, Reader in Mechanical Engineering, Imperial College of Science and Technology, University of London
DIESEL ENGINEERING DESIGN AND APPLICATIONS
S. D. HADDAD, Associate Professor, Western Michigan University, USA, and Director of HTCS Co., UK. and N. WATSON, Reader in Mechanical Engineering, Imperial College of Science and Technology, University of London
TECHNIQUES OF FINITE ELEMENTS
BRUCE M. IRONS, University of Calgary, and S. AHMAD, Bangladesh University, Dacca
FINITE ELEMENT PRIMER
BRUCE IRONS and N. SHRIVE, University of Calgary
CONTROL OF FLUID POWER: ANALYSIS AND DESIGN 2nd (Revised) Edition
D. McCLOY, Ulster Polytechnic, N. Ireland and H. R. MARTIN, University of Waterloo, Ontario, Canada
UNSTEADY FLUID FLOW
R. PARKER, University College, Swansea
DYNAMICS OF MECHANICAL SYSTEMS 2nd Edition
J. M. PRENTIS, University of Cambridge
ENERGY METHODS IN VIBRATION ANALYSIS
T. H. RICHARDS, University of Aston Birmingham
ENERGY METHODS IN STRESS ANALYSIS: With Intro. to Finite Element Techniques
T. H. RICHARDS, University of Aston in Birmingham
COMPUTATIONAL METHODS IN STRUCTURAL AND CONTINUUM MECHANICS
C. T. F. ROSS, Portsmouth Polytechnic
FINITE ELEMENT PROGRAMS FOR AXISYMMETRIC PROBLEMS IN ENGINEERING
C. T. F. ROSS, Portsmouth Polytechnic
ENGINEERING DESIGN FOR PERFORMANCE
K. SHERWIN, Liverpool University
ROBOTS AND TELECHIRS
M. W. THRING, Queen Mary College, University of London
MECHANICAL VIBRATIONS WITH APPLICATIONS
A. C. WALSHAW, Professor Emeritus, The University of Aston in Birmingham

COMPUTER-AIDED ENGINEERING
Heat Transfer and Fluid Flow

A.D. GOSMAN, B.A.Sc., Ph.D.,
B.E. LAUNDER, D.Sc (Lond), Sc.D. (M.I.T.),
G.J. REECE, B.A., M.Sc., D.I.C., Ph.D.

ELLIS HORWOOD LIMITED
Publishers · Chichester

Halsted Press: a division of
JOHN WILEY & SONS
New York · Brisbane · Toronto

First published in 1985 by
ELLIS HORWOOD LIMITED
Market Cross House, Cooper Street, Chichester, West Sussex, PO19 1EB, England

The publisher's colophon is reproduced from James Gillison's drawing of the ancient Market Cross, Chichester.

Distributors:

Australia, New Zealand, South-east Asia:
Jacaranda-Wiley Ltd., Jacaranda Press,
JOHN WILEY & SONS INC.,
G.P.O. Box 859, Brisbane, Queensland 4001, Australia

Canada
JOHN WILEY & SONS CANADA LIMITED
22 Worcester Road, Rexdale, Ontario, Canada.

Europe, Africa:
JOHN WILEY & SONS LIMITED
Baffins Lane, Chichester, West Sussex, England.

North and South America and the rest of the world:
Halsted Press: a division of
JOHN WILEY & SONS
605 Third Avenue, New York, N.Y. 10158, U.S.A.

British Library Cataloguing in Publication Data

Gosman, A.D.
Computer-aided engineering: heat transfer and fluid flow.
(Ellis Horwood series in mechanical engineering).
1. Fluid dynamics – Data processing
2. Heat – Transmission – Data processing
I. Title, II. Launder, B.E., III. Reece, G.T.
532'.051 QA911

ISBN 0-85312-866-9 (Ellis Horwood Library edition)
ISBN 0-85312-868-5 (Ellis Horwood student edition)
ISBN 0-470-20212-2 (Halsted Press edition)

Printed in Great Britain by Acfords, Chichester

Preface

Leading R & D centres now tackle dauntingly difficult problems of heat and fluid flow by computer modelling. Fine details of highly complex phenomena are simulated in this way: combustion of the fuel-air mixture in an IC engine; the ablation of the heat shield on a space re-entry vehicle, or the pulsatile flow of blood in an artery. Yet, by contrast, particularly at first-degree level, the computer has had far less impact on how the subjects of fluid mechanics and heat transfer are taught. True, more emphasis is now placed on numerical methods and computer modelling, but development of a conceptual understanding of the subjects is nearly always by traditional methods: a verbal exposition of the underlying physics by the teacher, assisted by analytical solutions of simple cases and by laboratory experiments.

While we should not wish to discard any of these traditional aids to learning it seemed to us that they could be strengthened by the more effective use of the computer in an instructional mode. With the help of a grant from the National Development Programme in Computer Assisted Learning we embarked on the development of a final-year computer-based course in fluid mechanics and heat transfer in which the main role of the computer was to reinforce the student's grasp of the underlying physics – to help him or (alas only rarely) her to learn what the equations mean. The course was centred on a specially-written master program, TEACH–C, for solving the generalised time-dependent Poisson equation. The code was structured so as to facilitate its adaptation from one physical situation to another. In this way all the heat transfer and fluid flow problems considered in the course could be tackled by making fairly minor changes to TEACH–C. How the separate case studies are organised for class study is described in the first section of this book. It was our deliberate intention that the class should be able to run and adapt the program and learn from the results it produced without having to understand the inner workings of TEACH–C. Nevertheless the TEACH–C solution procedure is organised sufficiently simply that in practice natural curiosity led most of our class to a comfortable familiarity with the code within a month or so of first using it.

After several years of in-house testing and refinement the most highly-developed of the case studies were released for class use at a number of institutions in the UK and the USA. Following the feedback we received, a selection of the problems was re-edited and brought together to form the present book.

Perhaps a few words of orientation are in order for those contemplating buying this book or who, having already done so, are considering how to use it. Firstly, the book is intended as a companion to rather than as a replacement for heat-transfer textbooks of the traditional kind. The teacher must decide what balance to strike between analytical and computational material. Even if the analytical examples we provide are sufficient for his purposes he will almost certainly need to give more background on their origin. A few of the early case studies may appear rather elementary.

Our experiences suggest that their depth is about right, however, for the start of the course: at any rate if the mode of learning is new to the student.

In postgraduate courses, students having completed the case studies in the book could, under the instructor's guidance – or perhaps on their own initiative – try adapting the code to simulate new physical phenomena or boundary conditions. For example, the introduction via the source-term facility in the program of terms representing the convective transport of heat (with a prescribed velocity field) provides entry to rich veins of fundamental and practical issues. The effect of conduction in the wall on apparent heat-transfer coefficients in pipe flow and the optimum siting of internal cooling holes in a turbine blade are just two problems successfully tackled as mini-projects within such courses.

In this latter example the role of the computer has, of course, grown from being an aid in learning the physics to that of a problem solver. Such a progression is entirely natural; a student trained, through the case studies considered in this book, to look for the physics in the computer output, will be in a stronger position to identify blunders in his coding than one who jumps straight in. For this reason we have also found that that the material in this book provides a useful initial orientation for research students on computational heat and fluid flow with little or no experience in the area. Such students will generally work through the material in a self-taught, self-paced mode. Several such students have gone on to publish papers in the scientific literature based on their further explorations with TEACH–C. We look forward to hearing from readers who are aiming to do likewise.

We have already mentioned the support of the Council for Educational Technology, through the NDPCAL, without which our work could never have got off the ground. A major contribution to the effort involved was provided by two colleagues: Dr F.C. Lockwood, who created the initial versions of several examples and Dr P.A. Newton, who felicitously transformed ideas and formulae into easily readable FORTRAN. Sincere thanks are also owed to our undergraduate students who worked enthusiastically with, in some cases, early draft versions of the material, and who succeeded in spite of the errors and ambiguities the documents contained. Every member of the first class we taught deserves a medal. Two of them, A.O. Demuren and I.K. Jennions, even stayed on through the summer after their graduation to help sort out some of the difficulties in the software before the next class arrived. Among those who have helped in the final editing stages we especially wish to thank Mr B.J. Dinsley for ensuring compatibility of the software with FORTRAN 77 and for arranging the provision of most of the UPDATE listings that appear in the book.

August 1985

A.D. Gosman
B.E. Launder
G.J. Reece

SOFTWARE
The authors have prepared the TEACH-C software for main frame computers. If there is sufficient demand, it could be made available for running on certain popular microcomputers and, if persons requiring such programs will send details of their own requirements and machine specifications, the authors will endeavour to provide a version compatible with that machine. Please write for further information to Ellis Horwood Ltd., Market Cross House, Cooper Street, Chichester PO19 1EB, England.

Table of contents

TEACH–C

User's Guide and Instruction Manual

---

1. Introduction

1.1 Outline of capabilities of TEACH–C

The computer program TEACH–C* provides a finite-difference solving procedure for the partial differential equation

$$\rho c_v \frac{\partial T}{\partial t} - \nabla \cdot (k \nabla T) - s = 0 \qquad (0.1)$$

which describes the conduction of heat; T is temperature, c_v specific heat, t time, k thermal conductivity and s a source or sink of energy. The program, written in FORTRAN 77, has been devised as a teaching aid to help undergraduate (or postgraduate) students

- to understand the physical behaviour of heat conduction and analogous phenomena under a wide range of conditions, and
- to discover consequences of practical interest.

In constructing the program, great emphasis has been laid on generality and adaptability. These qualities enable the user of the program, by performing only minor modifications, to simulate vastly different problems of heat transport, even including certain problems of three-dimensional heat transfer.

There is a wide range of other interesting phenomena in the applied sciences which can be described by the same type of equation as (0.1) or some equation derived from it. For example, replacement of the temperature T by the concentration C yields, with appropriate changes to the material properties, the transport equation for a diffusing species in a stationary medium.

Within the field of fluid mechanics, the stream function ψ in an irrotational flow obeys a degenerate form of equation (0.1), namely

$$\nabla^2 \psi = 0 \qquad (0.2)$$

---

* A listing of TEACH–C is provided in Appendix C.

while the distribution, in a fluid of viscosity μ, of axial velocity W in a fully-developed laminar duct flow, with axial pressure gradient $\mathrm{d}p/\mathrm{d}z$ is governed by another degenerate form of equation (0.1):

$$\nabla \cdot (\mu \nabla W) + \frac{\mathrm{d}p}{\mathrm{d}z} = 0 \qquad (0.3)$$

Additional examples include flows in porous media, pipe networks and lubricant films.

It is evident therefore that there is a wide range of phenomena governed by equations like (0.1) – (0.3) which can be solved by application of the TEACH–C program. Indeed the differences in the governing equations lie purely in the names conventionally attached to the dependent variables, in the presence or absence of individual terms, and in the types of initial and boundary conditions imposed. TEACH–C is designed to allow these variations to be incorporated at will.

A further, and somewhat different, area in which TEACH–C can contribute to teaching is in the field of numerical analysis. The emphasis on generality and adaptability has led to a program structure into which different finite-difference approximations or different iterative practices may easily be incorporated. Users may thus discover for themselves which combination of schemes is best in given circumstances, from the points of view of numerical accuracy, stability and speed of execution.

Finally, it should be mentioned that TEACH–C is the simplest member of a family of computer programs (the TEACH "family") for the analysis of fluid flow and heat transfer in two and three dimensions [ref. (0.1) and ref. (0.2)]. The other members of the family are, in all senses, substantially larger programs than TEACH–C for they provide solving schemes for the *simultaneous* solution of both the fluid flow and heat transfer phenomena. Such programs are used extensively in teaching and research at Imperial College London and many other institutions in Europe and the United States. It is our experience that a period of apprenticeship in learning TEACH–C and how to adapt it to solve various problems provides a valuable preparation for using one of the more 'senior' members of the TEACH family.

1.2 Purpose and Layout of the Guide

The present document is intended to provide a guide to TEACH–C including the physical and mathematical model on which it is based. *Section 2* presents the general form of the partial differential equation considered and provides a derivation of the set of algebraic "difference" equations which are used to characterise eqn. (0.1) in TEACH–C. Also provided are a description of the method of solution of the difference equations, the treatment of boundary conditions and such topics as convergence, stability and accuracy. *Section 3* is concerned with how the solving procedure described in Section 2 is translated into a computer program. First the overall structure is discussed and the most important conventions and FORTRAN symbols are introduced. A description is then provided of the application of the TEACH–C program to an example problem: the subroutines are considered in turn and the important operations performed by each are discussed. *Appendices* provide a glossary of FORTRAN notation, a listing of the computer program and a sample output, together with information on the TEACH 'PROBLEM's.

1.3 A Method of Using TEACH–C in a CAL Course

Precisely what, in any given context, represents the best way of using TEACH–C depends on numerous factors: the computer facilities available, the various aspects the teacher wishes to emphasise, the academic level of the students, the length of the course, etc. Clearly no general rules can be given. It may, however, be helpful to describe briefly the way that the program has been used by staff in the Mechanical Engineering Department at Imperial College because this has shaped the format of the supporting documentation prepared for use in conjunction with this Guide. The main features of the teaching and computing strategy are summarised in the following paragraphs; more detailed accounts have been published in references [0.3] and [0.4].

The Course. The material associated with TEACH–C has been used in a final year undergraduate course on fluid flow and heat transfer. The course begins with a derivation of the governing set of partial differential equations. Next, the material given in Sections 2 and 3 of this Guide is presented, about two lecture hours being devoted to this activity. We expect students to understand (and, given time, to be able to derive) the method by which the partial differential equation is converted to a set of difference equations. We require only a general knowledge of the program structure, however, and familiarity with the meaning of the more important variable names. Indeed, a student can perform all the computer exercises without detailed knowledge of the computer program, as will be described below.

After the program has been introduced to the class, the remaining class periods are shared roughly equally between conventional teaching activities and computer-related activities. For each particular topic considered (e.g., two-dimensional heat conduction processes, fully-developed duct flows) a typical work unit would consist of:

'conventional' activities	1 or 2	50-minute teaching period(s) discussing analytical approaches to the topic, describing the governing parameters and the observed behaviour;
	½	period discussing with students their analytical solutions of assigned problems;

computer – related activities	½	period discussing with students their computer solutions of assigned problems;
	1	period in which students report their discoveries from the computer solutions to the rest of the class.

In addition, students are expected to perform some of the computer explorations in their own time.

Concerning the computer-related activities, we attach particular importance to the student reporting sessions, for two reasons: firstly, they enable us to subdivide the class into small groups, then assign different computing tasks to each group and ensure that their results are made known to the whole of the class; and secondly, the students gain valuable experience of oral presentation, which noticeably improves their abilities in this respect.

It is clear that the coverage of any particular topic, as described by the above work-scheme, is rather brief. Twice as much time could easily have been devoted to each subject, though, in the course as given at Imperial College, the possibility of greater depth of treatment was ruled out by the breadth of subject matter we decided to cover. The decisions as to how much of the TEACH–C material should be incorporated into any course, what role it should play, and at what speed it should be covered, are ones that must be taken by each teacher intending to use the material. That choice is made easier, we trust, by the structure of the material presented: especially insofar as the set of computer exercises for each topic is treated independently of the rest and can thus be selected for extra emphasis or for omission without necessary detriment to the rest of the material.

Computing strategy. Although TEACH–C is a general-purpose code, the changes required to adapt it to particular applications may sometimes be extensive. In a teaching context this may place excessive demands on student time. To avoid this, we prepared in advance the necessary modifications for selected applications (details of which are given below) and arranging for them to be stored in a computer file which could be readily invoked by the students.

This strategy was facilitated by the availability of the UPDATE editing system on the CDC machines which we used.* UPDATE allows a user to create and run a modified version of a program held on file within the computer merely by submitting a card deck of the modifications (an UPDATE deck) prefixed by appropriate control cards. In fact, the UPDATE may itself be held on file to be called into action merely by the submission of the control cards. Moreover, on CDC computers any number of UPDATE modification files may be assembled in series; any particular UPDATE can prescribe modifications either to the basic program or to the UPDATE files already assembled.

* Facilities similar to UPDATE are available on other makes of computer. A description of UPDATE is provided in an appendix to the PROBLEM booklets of this series (see Appendix E below). The description is intended to serve also as a complete set of instructions for the manual implementation of the scheme if no automatic method is available.

We have found it convenient to work with three levels of UPDATE as shown in fig.0.1a. The first level modifies TEACH–C so that it is set up to study a particular topic or 'PROBLEM'. This is referred to as a PROBLEM UPDATE. The second level, which creates a 'LESSON' and is termed a LESSON UPDATE, modifies TEACH–C and the PROBLEM UPDATE so that a particular situation within the general class is simulated. Both PROBLEM and LESSON UPDATES are held on file in the computer; students merely invoke these files, they do not create them. The final level, a RUN UPDATE, consists of a few cards prepared by the student to explore the effects of specific parameters on the computed behaviour.

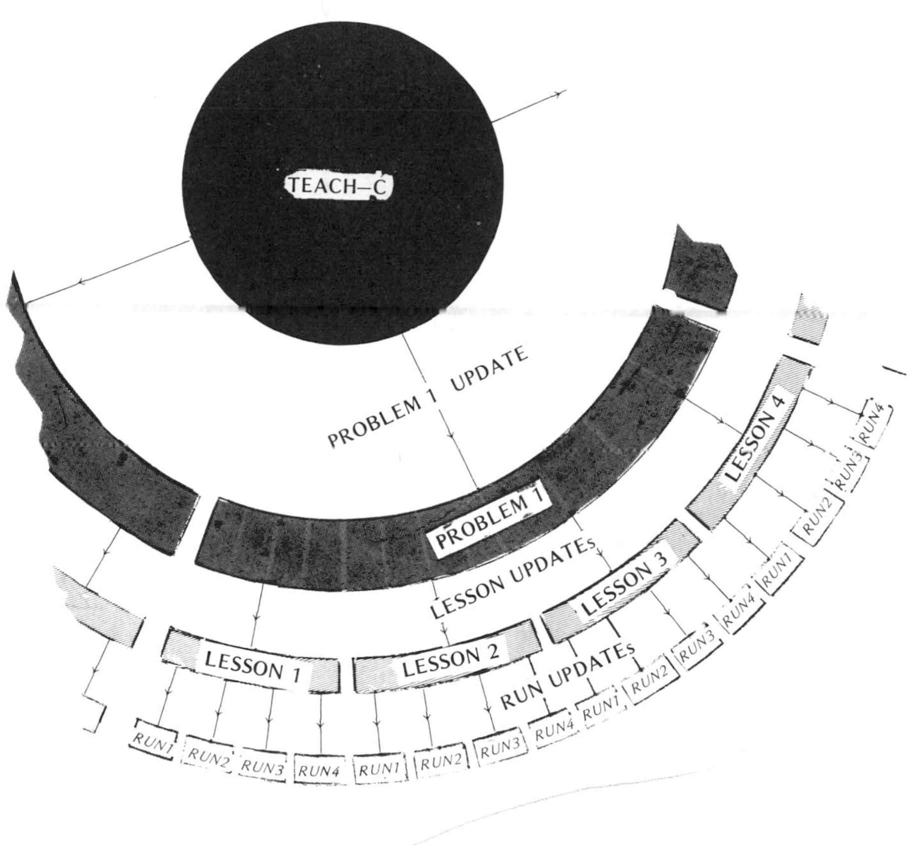

Fig. 0.1a The structure of the TEACH–C UPDATE scheme

PROBLEM	LESSON	RUN
1 **Unsteady one-dimensional heat conduction**	1 *Plane slab with step change in surface temperature*	Material properties of: 1 steel 2 copper
	2 *Effect of thermal resistance in the fluid layer*	1, 2, 3 Vary Biot number by changing heat-transfer coefficient
	3 *Conduction with a temperature-dependent heat source*	1, 2, 3 Vary heat-source strength 4, 5 Vary heat-transfer coefficient
	4 *Conduction with a sinusoidally-varying surface temperature*	1, 2 Vary period of fluctuations 3 Use moon soil data 4, 5, 6 Vary material properties.

Fig. 0.1b The structure of a typical TEACH–C PROBLEM

It should be emphasised that the UPDATE facility is a convenience but *not* a requirement for using TEACH–C in basically the way we have described. When UPDATE is not available it will usually be desirable to hold composite binary files corresponding to each LESSON (i.e., a file consisting of TEACH–C with the modifications indicated by the appropriate PROBLEM and LESSON UPDATE files actually included). Additionally minor adjustments would be needed so that the student could make the parameter explorations by way of DATA cards. Such an arrangement may also be desirable when there is great need to economise on computer time: the composite files may be held in binary form thus saving compilation time with each run.

Documentation. A series of student workbooks (see Appendix E) has been prepared for use in conjunction with this Guide. Each document is concerned with a particular PROBLEM. It provides a short introduction to the topic, a description of the changes needed to modify TEACH–C followed by a listing of the PROBLEM UPDATE, and a series of specific LESSONs.

Each LESSON commences with a description and listing of the LESSON UPDATE, followed by suggestions for parametric explorations and the necessary RUN UPDATE to effect them. Students are asked to plot various aspects of their computed results on graphs with suitably-chosen scales which are included * in the PROBLEM workbooks and questions are posed about the computed behaviour. A list of the LESSONs comprising a sample PROBLEM is given in fig. 0.1b.

Of course if suitable machine plotting facilities (including provision for hard copies) are available to the user it would be sensible to exploit them. The wide variations and incompatibilities in plotting software and hardware have prevented us from incorporating this mode of presentation as a standard feature in the code.

2. Mathematical Foundations

2.0 *Introductory remarks*

It was mentioned in the Introductory section that although TEACH–C is identified with heat conduction, it can in fact be regarded as a general-purpose solver of a form of partial differential equation which arises in a wide variety of physical situations. For the sake of brevity and concreteness, however, the descriptions of the method of analysis and computer program contained in what follows will focus on heat conduction only, in the knowledge that the PROBLEM workbooks provide ample examples of the adaptation of the program to other phenomena.

2.1 *Differential equation of heat conduction*

We are here concerned with conduction in a stationary medium, in circumstances in which the temperature T may vary with time t, and just two spatial co-ordinates, which for plane configurations will be taken as the Cartesian directions x and y, while for bodies of revolution the cylindrical-polar system will be used, with x now denoting the axial co-ordinate and r the radial one.

Application of the fundamental laws of heat conduction yields a differential equation for T, whose detailed derivation is given in many textbooks, for example refs. [0.13] and [0.14]. Here the result will simply be quoted, in terms of cylindrical-polar co-ordinates:

$$\rho c_v r \frac{\partial T}{\partial t} - \frac{\partial}{\partial x}\left\{rk\frac{\partial T}{\partial x}\right\} - \frac{\partial}{\partial r}\left\{rk\frac{\partial T}{\partial r}\right\} - rs = 0 \qquad (0.4)$$

* These were found to enhance considerably the quality of the presentations in the 'report-back' sessions.

where

$\rho \equiv$ density of the medium
$c_v \equiv$ specific heat at constant volume
$k \equiv$ thermal conductivity
$s \equiv$ distributed sources or sinks of energy

The Cartesian version of eqn. (0.4) is easily obtained by setting r equal to unity wherever it does not appear in a differential coefficient and replacing it by y elsewhere.

Auxiliary information. In order to specify completely a particular heat conduction problem in mathematical terms, it is necessary to provide the following additional information:

(*a*) values of the material properties, ρ, c_v and k, and the distributed source s (or the appropriate functions, if any of these should depend on time, position or temperature);
(*b*) the initial distribution of temperature throughout the medium (strictly necessary only for time-dependent problems);
(*c*) prescriptions on the temperatures and/or heat fluxes at all locations on the boundaries of the medium, at all instants of time. These prescriptions have the general form:

$$f_1 \frac{\partial T}{\partial n} + f_2 T + f_3 = 0 \tag{0.5}$$

where n denotes the outward normal to the boundary and f_1, f_2 and f_3 are functions which specify the particular condition which applies. Table 0.1 below gives the definitions of these functions for some of the more commonly encountered boundary conditions:

TABLE 0.1 Definitions of f_1, f_2 and f_3 in eqn. (0.5) for various common boundary prescriptions.

Nature of boundary condition	f_1	f_2	f_3
Prescribed surface temperature T_B	0	1	$-T_B$
Prescribed surface heat flux $\dot{q}''_B$	k	0	$-\dot{q}''_B$
Prescribed coefficient α of heat-transfer to an external flow at reference temperature T_F	k	α	$-\alpha T_F$

2.2 *Finite-difference equation*

(*a*) *Computational grid and associated notation*

For the purposes of computer solution a rectilinear grid, composed of co-ordinate lines, is imposed on the domain of interest and the temperatures are calculated at the discrete points formed by the grid intersections or *nodes*. The location and spacing of the lines is allowed to be arbitrary, so as to enable the nodes to be concentrated, and so enhance economy, in regions where temperature gradients are likely to be steep. Fig. 0.2 illustrates a typical grid superimposed on a hypothetical *T*-shaped body of revolution, which might, for example, be a component of a gas turbine.

Each grid node is imagined as being surrounded by its own control volume, or *cell,* with boundaries (shown as dashed lines on fig. 0.2) defined as the perpendicular bisectors of the grid lines connecting it to neighbouring nodes. Fig. 0.3 shows a typical node and its neigh-

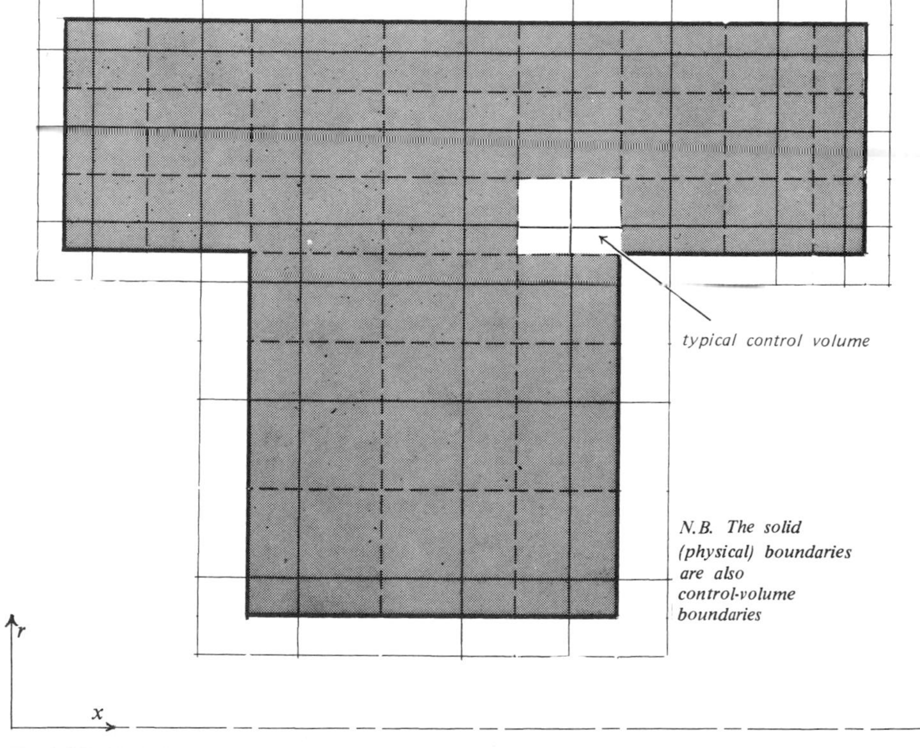

Fig. 0.2 A typical computational grid (*solid lines*) and associated control volumes (*dashed lines*)

bours, together with various items of notation: thus the central node of a cluster is labelled P, and its neighbours are labelled N, S, E and W in point-of compass fashion, and the intervening cell boundaries are denoted by lower-case versions of the same letters (see fig. 0.3). The co-ordinates of P are denoted by (x_P, r_P); the axial and radial dimensions of the enclosing cell by δx_{ew} and δr_{ns} respectively; and various inter-nodal distances such as δx_{PW}, δr_{NP} etc. are defined in a self-explanatory fashion in the figure.

Reference will also be made to the surface areas and volumes swept out by rotating the cells through one radius about the symmetry axis. The volumes will be denoted by V_P and approximated by:

$$V_P \approx r_P \, \delta x_{ew} \delta r_{ns} \tag{0.6}$$

and the areas by, for example:

$$a_n \approx r_n \delta x_{ew} \tag{0.7}$$

$$a_e \approx r_P \delta r_{ns} \tag{0.8}$$

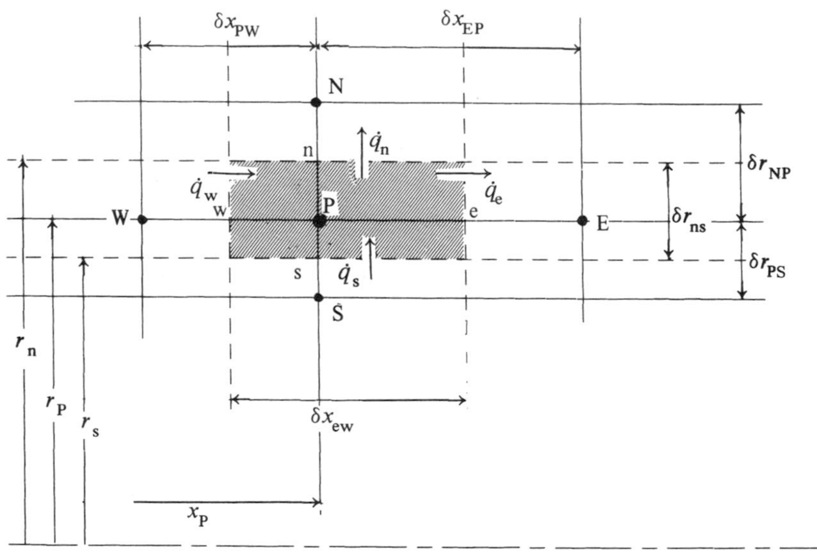

Fig. 0.3 A typical cluster of grid nodes and associated notation

(b) Derivation of equation

An algebraic, finite-difference counterpart of the differential equation (0.4) will now be derived for the representative cluster of grid nodes shown in fig. (0.3). The procedure will be as follows: firstly eqn. (0.4) will be integrated, as far as formal calculus allows, over the control volume surrounding P and simultaneously averaged over a finite increment of time δt; then the remaining integrals will be replaced by algebraic approximations.*

The starting point is therefore the integral expression:

$$\frac{1}{\delta t}\int_{t}^{t+\delta t}\int_{s}^{n}\int_{w}^{e}\Bigg[\underbrace{\rho c_v r\frac{\partial T}{\partial t}}_{I_1}-\underbrace{\frac{\partial}{\partial x}\left(rk\frac{\partial T}{\partial x}\right)}_{I_2}-\underbrace{\frac{\partial}{\partial r}\left(rk\frac{\partial T}{\partial r}\right)}_{I_3}-\underbrace{rs}_{I_4}\Bigg]\,\mathrm{d}x\,\mathrm{d}r\,\mathrm{d}t=0 \qquad (0.9)$$

The individual in terms I_1, I_2 etc. will now be evaluated, with assumptions and approximations introduced where necessary.

Thus, if we first assume that the material properties ρ and c_0 are constant with time and position (which is not essential, but is usually the case) the first integral may be written:

$$I_1=\rho\frac{c_v}{\delta t}\left[\int_{s}^{n}\int_{w}^{e} rT\,\mathrm{d}x\,\mathrm{d}r\right]_{t}^{t+\delta t} \qquad (0.10)$$

and then approximated by:

$$I_1\approx\rho c_v\frac{(T_P^{new}-T_P^{old})}{\delta t}V_P \qquad (0.11)$$

where the superscripts represent the new and old values at time $t+\delta t$ and t, respectively, and V_P is defined in eqn. (0.6).

The second integral can be partially evaluated:

$$I_2=-\frac{1}{\delta t}\int_{t}^{t+\delta t}\left\{\int_{s}^{n}\left[\left(rk\frac{\partial T}{\partial x}\right)\right]_{w}^{e}\mathrm{d}r\right\}\mathrm{d}t \qquad (0.12)$$

and then the following approximations are made: firstly, the contents of

---

* This procedure, it should be remarked, differs from the Taylor-series approach commonly found in standard textbooks (see, e.g. [0.15], [0.16]), although the results will be identical for identical assumptions. The present approach is preferred as having greater physical meaning..

the curly brackets are time-weighted thus:

$$\frac{1}{\delta t}\int_{t}^{t+\delta t}\left\{\qquad\right\}\mathrm{d}t \approx f\left\{\qquad\right\}^{\text{new}} + (1-f)\left\{\qquad\right\}^{\text{old}} \tag{0.13}$$

where f is a weighting factor lying in the range $0 \leqslant f \leqslant 1$, whose value will be specified later. Secondly, we recognise that the integrals within the curly brackets simply represent the total heat flows across the w and e faces of the cell. We might, therefore, reasonably approximate the flow across the w face, for example, by

$$\int_{\mathrm{s}}^{\mathrm{n}}\left[rk\frac{\partial T}{\partial x}\right]_{\mathrm{w}}\mathrm{d}r \approx r_{\mathrm{P}}\,\frac{(k_{\mathrm{W}}+k_{\mathrm{P}})}{2}\,\frac{(T_{\mathrm{P}}-T_{\mathrm{W}})}{\delta x_{\mathrm{PW}}}\,\delta r_{\mathrm{ns}} \tag{0.14}$$

which is a notional thermal-resistance formula for one-dimensional heat transfer between P and W. A similar formula can be deduced for the heat transfer between E and P. It may be noted that allowance has been made for (smooth) spatial variations in k by taking the arithmetic mean of the nodal values*, a practice which ensures no discontinuity of heat flow across a cell boundary and thereby preserves accuracy and physical realism. The third term I_3 in eqn. (0.6), for which the results of the integration will simply be given below, is treated similarly.

The evaluation of the final term I_4 will in general depend upon the particular form which s takes for the problem under consideration. It should always however be possible to express the result as a linearised and time-averaged function of T_{P} as follows:

$$I_4 = \frac{1}{\delta t}\int_{t}^{t+\delta t}\int_{\mathrm{s}}^{\mathrm{n}}\int_{\mathrm{w}}^{\mathrm{e}} r\,s\,\mathrm{d}x\,\mathrm{d}r\,\mathrm{d}t$$

$$\approx f\left(B_{\mathrm{P}}^{\text{new}}T_{\mathrm{P}}^{\text{new}} + C_{\mathrm{P}}^{\text{new}}\right) + (1-f)\left(B_{\mathrm{P}}^{\text{old}}T_{\mathrm{P}}^{\text{old}} + C_{\mathrm{P}}^{\text{old}}\right). \tag{0.15}$$

The formulation of I_4 in this way possesses certain advantages which will be discussed later.

Assembly of the foregoing results yields, in compact notation, the following finite-difference energy conservation equation:

$$D_{\mathrm{P}}(T_{\mathrm{P}}^{\text{new}} - T_{\mathrm{P}}^{\text{old}}) + f\left\{\sum_{c} A_{c}^{\text{new}}(T_{\mathrm{P}}^{\text{new}} - T_{c}^{\text{new}}) - (B_{\mathrm{P}}^{\text{new}}T_{\mathrm{P}}^{\text{new}} + C_{\mathrm{P}}^{\text{new}})\right\}$$

$$+ (1-f)\left\{\sum_{c} A_{c}^{\text{old}}(T_{\mathrm{P}}^{\text{old}} - T_{c}^{\text{old}} \qquad\qquad + C_{\mathrm{P}}^{\text{old}})\right\} = 0 \tag{0.16}$$

* Of course, if the variation is not smooth – e.g. stepwise – a different interpolation formula should be used.

where Σ_c denotes summation over the cluster of nodes N, S, E, W and the various coefficients have the following definitions and physical significance:

(*i*) $$D_P \equiv \rho c_v V_P / \delta t \qquad (0.17)$$

may be regarded as the *total thermal capacitance* of the volume.

(*ii*) The A_c's have the significance of *thermal conductances* (i.e. reciprocals of thermal resistance) for the heat flows between adjacent cells, as may be perceived from their definitions:

$$A_N \equiv (k_N + k_P)\, a_n / 2\delta r_{NP}$$

$$A_S \equiv (k_S + k_P)\, a_s / 2\delta r_{PS}$$

$$A_E \equiv (k_E + k_P)\, a_e / 2\delta x_{EP}$$

$$A_W \equiv (k_W + k_P)\, a_w / 2\delta x_{PW}$$

Specification of the weighting factor f. It is necessary at this stage to provide a specification for the temporal weighting factor *f*. Reference to numerical analysis textbooks (e.g. refs [0.15], [0.16]) will reveal that various practices have been used, each with its advantages and disadvantages. Two of the more popular practices will now be discussed. If *f* is specified as zero, eqn. (0.16) may be arranged to give the new temperatures T_P^{new} as an explicit function of the old values, as follows:

$$D_P T_P^{new} = \sum_c A_c^{old} T_c^{old} + (D_P - \sum_c A_c^{old} + B_P^{old})\, T_P^{old} + C_P^{old} \qquad (0.18)$$

This enables the T_P^{new} to be calculated by a simple application of eqn. (0.18) to each grid node. The drawback to this 'explicit' method, as it is sometimes termed, is that numerical stability requires that the contents of the coefficient of T_P^{old} in eqn. (0.18) should sum to a non-negative number, (for otherwise $T_P{}^{new}$ would not be bounded by T_P^{old} and the T_c^{old} when $s = 0$ as reality demands), implying a restriction on the maximum admissible time step. For the especially simple case of no sources or sinks, uniform conductivity and a uniform grid of spacing δ the restriction is

$$\frac{k(\delta t)_{max}}{\rho c_v \delta^2} \lesssim \frac{1}{4} \qquad (0.19)$$

which amounts to requiring that the temperature 'wave' should not propagate more than a cell dimension during the time interval. This is a severe restriction, which often renders this approach uneconomical, especially since $\delta t_{max} \propto \delta^2$ and a small value of δ is often dictated by accuracy requirements: accordingly the explicit scheme will not be used here.

It is shown, e.g., in ref. [0.16] that *no* time step restriction exists if f is chosen to lie in the range $0.5 \leqslant f \leqslant 1.0$ and that there are reasons for preferring $f = 0.5$, (known as the Crank-Nicholson scheme) at least for transient calculations. Such schemes are termed *implicit,* because T_P^{new} is now related to the surrounding 'new' temperatures T_c^{new}, thus requiring solution of a set of simultaneous equations. The latter requirement is however more than counterbalanced by the freedom to specify arbitrary δt, which is especially advantageous when only the steady-state solution is of interest. TEACH–C therefore uses an implicit scheme, with f set to unity so as to minimise demands on computer storage:* however, it can easily be modified to incorporate the Crank-Nicholson scheme, if desired. The equations solved by TEACH–C are therefore of the form:

$$D_P(T_P^{new} - T_P^{old}) + \sum_c A_c^{new}(T_P^{new} - T_c^{new}) - (B_P T_P^{new} + C_P^{new}) = 0 \tag{0.20}$$

From now on, for brevity, we shall simply drop the superscript 'new'.

(*c*) *Boundary conditions*

It is clearly important to ensure that the energy conservation principle is correctly applied throughout the entire domain of solution and this requirement, together with the definitions of the cells, has led to the practice illustrated in fig. 0.2, in which the grid has been located such that the physical boundaries coincide with the outermost boundaries of the adjoining cells. It is equally important to recognise that *the temperatures at the grid nodes external to the solution domain have no physical significance,* nor do the terms in eqn. (0.20) which relate to the external nodes; indeed it is precisely these terms which require replacement to incorporate the boundary conditions.

A convenient way of effecting replacement, from the computer-programming point of view, is first to suppress the redundant heat-flow term(s) in eqn. (0.20) by setting the relevant A_c to zero and then to insert the correct boundary expression as a *false heat source,* via the coefficients B_P and C_P. These practices will now be illustrated through the example of fig. 0.4, where the boundary in question, passing through B,

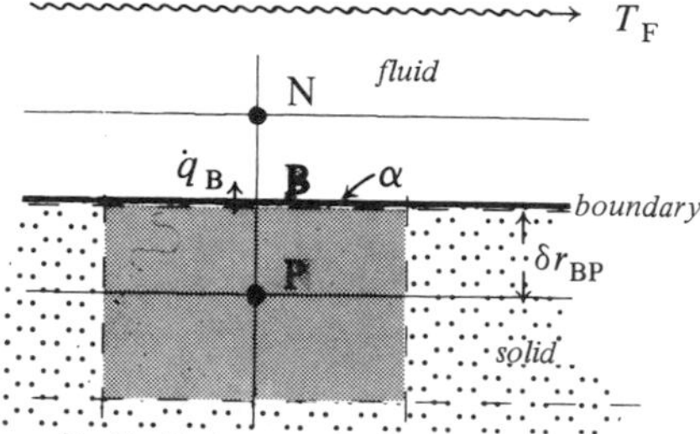

Fig. 0.4 Section of boundary exposed to a moving fluid

* A value less than unity can in any case give rise to oscillatory behaviour when δt exceeds a certain threshold value.

is engaged in convective heat transfer with a moving fluid at temperature T_F, and the heat transfer coefficient is α. The usual, and in this instance incorrect, expression for the heat flow $\dot{q}_N$ across the cell boundary in question is:

$$\dot{q}_N = -A_N(T_P - T_N)$$

which is suppressed by setting $A_N = 0$. Then, the correct heat flow is derived with the aid of eqn. (0.5) and the finite-differencing practices used earlier. It is

$$\dot{q}_N = -(T_P - T_F)/R \qquad (0.21)$$

where $R \equiv [2\,\delta r_{BP}/(k_B + k_P) + 1/\alpha]a_n$ is the total thermal resistance between node P and the fluid. Finally, the appropriate linearised-source coefficients are deduced to be:

$$B_P \equiv -1/R$$

$$C_P \equiv T_F/R$$

Table 0.2 summarises the linearised-source coefficients appropriate to the above and other common boundary conditions. Of particular note is the straightforward manner in which prescribed heat-flux conditions may be imposed by this method.

TABLE 0.2 Definitions of linearised-source coefficients appropriate to various common boundary conditions.

Type of boundary condition	B_P	C_P
Prescribed surface temperature T_B	$\frac{-k_{BP}a_n}{\delta r_{BP}}$	$\frac{k_{BP}a_nT_B}{\delta r_{BP}}$
Prescribed surface heat flux $\dot{q}''_B$	0	$\dot{q}''_Ba_n$
Prescribed coefficient α of heat transfer to external flow at temperature T_F	$-\frac{1}{R}$	$\frac{T_F}{R}$

Note: $k_{BP} \equiv \frac{1}{2}(k_B + k_P)$; $R \equiv [\delta r_{BP}/k_{BP} + 1/\alpha]\, a_n$

A further use of the false-source notion will now be described: it is sometimes desired that the temperature at some interior location in the grid be held at some prescribed value, say T_{fix}. This might, for example, be the temperature of the fluid in an internal cooling passage. Specification, at the node(s) in question, of:

$$\begin{aligned} B_{\text{P}} &\equiv -\gamma \\ C_{\text{P}} &\equiv \gamma\, T_{\text{fix}} \end{aligned} \tag{0.22}$$

where γ is a large number (e.g. 10^{30}) will cause the desired temperature to be imposed (to confirm this, substitute into eqn. (0.20) and solve for T_{P}, neglecting small terms).

2.3 Solution of difference equations

(a) *Substitution formula*

For the purposes of solution eqn. (0.20) is recast into the form:

$$(A_{\text{P}} - S_{\text{P}})\, T_{\text{P}} = \sum_c A_c T_c + S_{\text{U}} \tag{0.23}$$

where:

$$\begin{aligned} A_{\text{P}} &\equiv \sum_c A_c \\ S_{\text{U}} &\equiv C_{\text{P}} + D_{\text{P}} T_{\text{P}}^{\text{old}} \\ S_{\text{P}} &\equiv B_{\text{P}} - D_{\text{P}} \end{aligned}$$

One such equation can be written for each interior cell, yielding a set of simultaneous algebraic equations, whose number equals that of the unknown temperatures: the task is now therefore to solve the simultaneous equations.

There exist a variety of solution procedures which might be used, ranging in complexity from simple point-iteration methods to direct matrix-inversion techniques: details may be found in refs. [0.15] and [0.16] among others. The structure of the TEACH program is such as to admit any of the available procedures, but we have chosen not to employ in the standard version either of those just mentioned. We use instead a combination of *iteration by lines* and *block adjustments,* the aim being to achieve computational efficiency without excessive complication and computer storage. These procedures will now be described.

(*b*) *The line-iteration procedure*

This procedure involves simultaneous solution for the temperatures along each grid line, while the temperatures along neighbouring lines are temporarily taken as 'known', the most recently-calculated values being used: it may therefore be thought of as the line-by-line counterpart of point Gauss-Seidel iteration (ref. [0.15]). The simultaneous solution is achieved by a particular form of Gaussian elimination known variously as the 'Thomas Algorithm' (ref. [0.17]) or 'Tri-Diagonal Matrix Algorithm (TDMA)'.

The line-by-line procedure is applied along north-south grid lines, starting at the westmost one and sweeping eastwards. For each line, eqn. (0.23) is written as:

$$(A_P - S_P)\,T_P = A_N T_N + A_S T_S + S'_U$$

where $S'_U \equiv S_U + A_E T_E^{N-1} + A_W T_W$ is evaluated from the currently-available temperatures. Here the superscript $^{N-1}$ is appended to the T_E's to indicate that they have the values from the $(N-1)$th iteration while the rest are at level N. The set of equations for each line therefore has the form:

$$d_j T_j = a_j T_{j+1} + b_j T_{j-1} + c_j \qquad (0.25)$$

where j denotes the position along the line.

The TDMA allows eqn. (0.25) to be replaced by the back-substitution formula:

$$T_j = a'_j T_{j+1} + c'_j \qquad (0.26)$$

whose coefficients are defined by:

$$a'_j \equiv a_j/(d_j - b_j a'_{j-1})$$

$$c'_j \equiv (b_j c'_{j-1} + c_j)/(d_j - b_j a'_{j-1})$$

and whose origins are explained in Appendix B. The procedure for application is first to assemble the a'_j and c'_j, proceeding in order of increasing j to the uppermost value J, say. Then the T_j are obtained from eqn. (0.26) in descending order of j, aided by the fact that $a'_J = 0$.* One sweep of the entire set of grid lines represents a cycle of line iteration.

(*c*) *The block adjustment procedure*

The action of an iterative procedure is to sweep the errors in the prevailing solution of the temperature field to the boundaries, where they are reduced or eliminated by the boundary conditions. That the errors are not reduced to zero in just one iteration is revealed by consideration of the residual sources R_P of the finite-difference equations, defined by:

* A consequence of the incorporation of the boundary conditions by manipulating the finite-difference coefficients in the manner described in §2.2(*c*), which automatically produces the correct TDMA coefficients.

$$R_P \equiv (A_P - S_P)\,T_P - \sum_c A_c T_c - S_U \tag{0.27}$$

Physically, the R_P's represent the net energy imbalances for the cells, which should of course ideally be zero – i.e. if the prevailing temperature field were without errors.

Now, by definition, the application of the TDMA to a grid line reduces the R_P's at all nodes along that line to zero: however, since they depend on temperatures on neighbouring lines, they become finite again (but usually smaller than in the previous iteration) when the neighbouring temperatures are adjusted.

Unfortunately there are circumstances in which the rate of reduction of the R_P's by the line-iteration procedure becomes unacceptably slow: one such instance is when the resistance to heat transfer at the boundaries is large, as may happen, for example, in convective cooling when the heat transfer coefficient α is small* (inspection of eqn. (0.21) will confirm this). Since it is the changes in boundary heat flow which effectively remove the errors, the effect of the high resistance is to cause them to be reflected back into the interior field, without significant reduction.

An effective remedy for this problem is to invoke a procedure which, by simultaneously adjusting the temperatures on each line by uniform increments for each line, the value of the increment varying from line to line, causes the R_P's to sum to zero along every line, and hence over the entire field. (An alternative interpretation is that the adjustments cause the conservation of energy principle to be satisfied simultaneously over every vertical column of cells.)

The details of the procedure are as follows: let the prevailing solution produced by a line-iteration be denoted by $\widetilde{T}_{i,j}$ where i denotes the ith column (i.e. vertical grid line) and j the jth row (horizontal line). A uniform increment δT_i is to be added to the interior nodes on each column such that:

$$\begin{aligned}\sum_{\text{all } j} \Big\{ & (A^P_{ij} - S^P_{ij})(\widetilde{T}_{ij} + \delta T_i) - A^W_{ij}(\widetilde{T}_{i-1,j} + \delta T_{i-1}) \\ & - A^E_{ij}(\widetilde{T}_{i+1,j} + \delta T_{i+1}) - A^N_{ij}(\widetilde{T}_{i,j+1} + \delta T_i) \\ & - A^S_{ij}(\widetilde{T}_{i,j-1} + \delta T_i) - S^U_{ij} \Big\} = 0 \end{aligned} \tag{0.28}$$

where the subscripts of the parent eqn. (0.23) have been written as superscripts to avoid complication.

* Strictly, the correct criterion is the relative magnitudes of the convective and conductive resistances as measured by a *Biot number* $\text{Bi} \equiv \alpha L/k$, L being a typical dimension. The adverse situation is when Bi is small.

The following equation for the δT's results from further manipulation of eqn. (0.28):

$$d_i \delta T_i = a_i \delta T_{i+1} + b_i \delta T_{i-1} + c_i \tag{0.29}$$

where:

$$a_i \equiv \sum_j A_{ij}^{\mathrm{E}}$$

$$b_i \equiv \sum_j A_{ij}^{\mathrm{W}}$$

$$c_i \equiv - \sum_j \{(A_{ij}^{\mathrm{P}} - S_{ij}^{\mathrm{P}}) \widetilde{T}_{ij} - \sum_c A_{ij}^c \widetilde{T}^c - S_{ij}^{\mathrm{U}}\} = - \sum_j R_{ij}^{\mathrm{P}}$$

$$d_i \equiv \sum_j (A_{ij}^{\mathrm{E}} + A_{ij}^{\mathrm{W}} - S_{ij}^{\mathrm{P}})$$

Two features of the above equation are noteworthy: firstly, it will be seen that the 'source term' c_i for the temperature adjustments is nothing more than the residual source sum, $\Sigma R_{ij}^{\mathrm{P}}$, as would be expected from the earlier discussion. Secondly, the equation is of similar form to eqn. (0.25) and is therefore amenable to solution by the TDMA.

(*d*) *Outline of complete procedure*

Fig. (0.5) is a flow chart of the solution procedure, in which are shown the various steps involved in advancing the temperature field through time, starting from the initial distribution at time $t = t_0$. Thus, the procedure advances to the next time level $t = t_0 + \delta t$ by first evaluating the coefficients of the difference equations, basing them on the prevailing temperatures as an initial estimate. Then, the temperatures are updated by first performing a line-iteration sweep and then making the block adjustments. The result so obtained is examined for satisfactoriness, by criteria described below. If it is unsatisfactory, the cycle is repeated from the coefficient-calculation stage, until an acceptable solution is obtained. Then, unless the desired period of time has been covered, the method proceeds to the next time level, storing the newly-obtained temperature field as 'old' values.

A variant on the above procedure is employed when only the 'steady-state' solution is of interest. In these circumstances, it is advantageous to perform just one 'time' step, with δt effectively set equal to infinity (the use of the implicit formulation allows this, to great advantage). The solution is then obtained simply by iterating to completion.

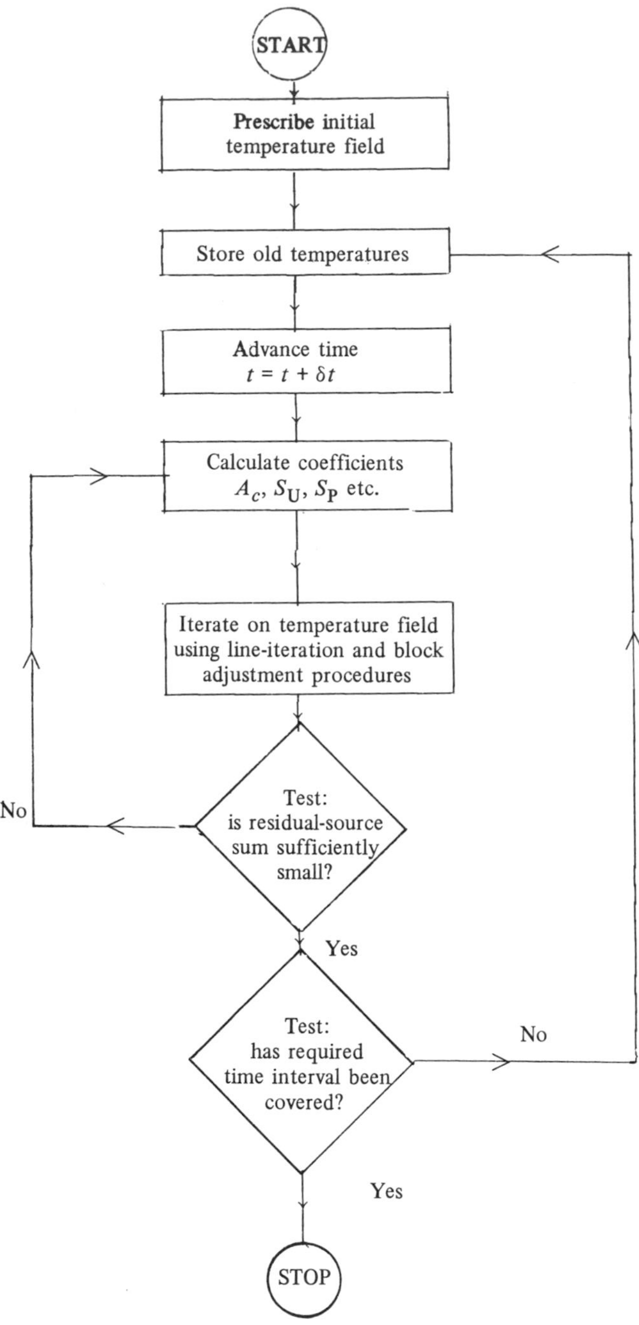

Fig. 0.5 Flow chart of the solution procedure

(*e*) *Miscellaneous matters*

Tests for convergence: 'Convergence' is defined as the property of the iteration scheme to proceed smoothly from a set of initial values to an acceptable solution of the finite-difference equations, where acceptability is measured by

(*i*) reduction in the changes in temperature produced by successive sweeps to a prescribed low value.
(*ii*) reduction of the 'residual sources' R_P of the finite-difference energy-conservation equations, defined by eqn. (0.27), to prescribed small values.

Particular emphasis is placed on the latter criterion, and it is usually required that the sum of the absolute values of all residual energy sources in the field should be less than a specified small fraction of a suitable reference quantity, such as the total heat transfer rate.

Accuracy: In common with all finite-difference procedures, the accuracy of the present procedure depends not only on obtaining an acceptable solution to the difference equations, but also on specifying sufficiently small space and time intervals: thus it is necessary always to ensure that the solution is invariant to further reduction in these intervals.

Numerical stability: The finite-difference eqn. (0.23) has been formulated to satisfy as far as possible a standard requirement for convergence of an iterative solution method, namely that of diagonal dominance of the coefficient matrix (ref. [0.17]). In mathematical terms, this requirement can be expressed as:

$$|A_P - S_P| \geqslant \sum_c |A_c| \tag{0.30}$$

at all grid cells, with strict inequality for at least one cell. It may be confirmed that, in the absence of heat sources which increase with temperature (for which $B_P > 0$) this criterion is always obeyed, irrespective of the grid and time intervals; while for the special case just mentioned, obedience may normally be procured by appropriate specification of these intervals, provided that the problem posed is a physically realistic one.

The foregoing criterion is strictly valid only for linear equations with constant coefficients: fortunately, however, it is a sufficient rather than a necessary condition; and experience has shown it to be useful even when the coefficients are (mild) functions of temperature.

For strongly non-linear problems, it is sometimes necessary to damp the changes in temperature which occur between iterations. There are two means available for achieving this: one is to employ small time steps and the other is to introduce under-relaxation, by evaluating T_P as

$$T_P = \beta T_P{}^N + (1 - \beta) T_P{}^{N-1} \tag{0.31}$$

where T_P^N and T_P^{N-1} are the solutions at the Nth and $(N-1)$th iterations, respectively, and β is the under-relaxation factor, having a value between zero and unity. This feature may be implicitly incorporated into the procedure by substituting for T_P^N in eqn. (0.31) from eqn. (0.23) to yield:

$$[(A_P - S_P)/\beta]\ T_P = \sum_c A_c T_c + S_U + (1-\beta)\, T_P^{N-1}(A_P - S_P)/\beta \quad (0.32)$$

3. Description of the computer program

3.1 Introduction

3.1.1 *Capabilities and limitations*

TEACH–C has the same breadth of applicability as the solution method described in Section 2 above. The program is therefore applicable to situations which are transient or steady-state, plane or axisymmetric, with uniform or variable thermal conductivity and with distributed heat sources. No restrictions are imposed on the form of, or conditions imposed at, the boundaries, although composite rectangular shapes can be handled with greatest ease.

The program may be altered with little difficulty to allow for variations in density or specific heat or for non-isotropic thermal conductivity. However, for reasons which will be explained below, we have not included these features, which would entail additional computer storage, as built-in options.

The programming language is FORTRAN–IV and the program has been developed and tested on CDC 6400 and 6600 series machines: it has also been successfully used on a variety of other makes of machines.*

TEACH–C, as all computer programs, represents a compromise between the often-conflicting demands of generality, efficiency (as measured by computer time and storage requirements) and ease of understanding. We regard all these considerations as important, but an overriding requirement has been to minimise demands on computer storage, even though this may sometimes be at the expense of longer computing times. We did this in the knowledge that many potential users would wish to mount the program on small computers with limited storage. However, those users who are not so restricted, may, with very little effort, reduce running times in ways which will be described later.

3.1.2 *Introduction of the bi-radial coordinate frame.*

One device which we have employed to reduce computer storage is to store the coefficients of the finite-difference equations in one-dimensional arrays, which are overwritten as the program proceeds from one line to the next. This arrangement, coupled with the method of solution described in Section 2.3 above, could impose a certain inflexibility on the program

* TEACH–C requires 12 000 (60-bit) words to be loaded on the CDC 6400.

which, if not removed, might in some cases be disadvantageous. A case in point is the annular fin illustrated in fig. 0.5 below, where it is better to apply the solving procedure along lines which run perpendicular to the predominantly-radial heat flow. If the fin were not axially symmetric, one would simply reorient it, but this is not allowed within the context of the standard cylindrical-polar frame (*a*). It *is* however possible if the "bi-radial" version (*b*) is employed, wherein an alternative symmetry axis is introduced, parallel to the y co-ordinate, so this is what we have done.

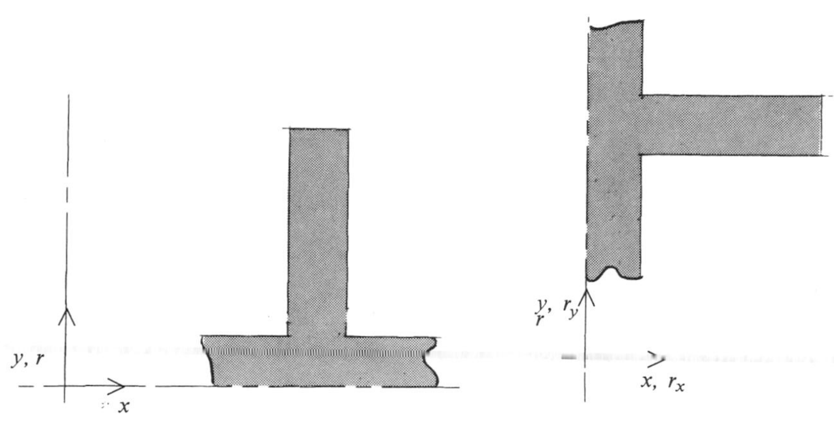

(*a*) Standard arrangement (*b*) Biradial arrangement, allowing alternative orientation

Fig. 0.6 Illustration of the use of a biradial coordinate frame for an axisymmetric problem

Thus there are now two radii of curvature, denoted in fig. 0.6 (b) by r_x and r_y respectively. Of course it is inadmissible ever to invoke more than one symmetry axis, so in practice the unwanted one is suppressed in the usual fashion by setting the relevant r to unity.

It is not difficult to generalise the finite-difference eqn. (0.16) to the biaxial co-ordinate frame. All that is necessary is to redefine the areas and volumes appearing in the coefficients as follows:

$$V_\mathrm{P} \approx r_{x,\mathrm{P}} r_{y,\mathrm{P}} \delta x_\mathrm{ew} \delta y_\mathrm{ns} \tag{0.33}$$

$$a_\mathrm{n} \approx r_{x,\mathrm{P}} r_{y,\mathrm{n}} \delta x_\mathrm{ew} \tag{0.34}$$

$$a_\mathrm{e} \approx r_{x,\mathrm{e}} r_{y,\mathrm{P}} \delta y_\mathrm{ns} \tag{0.35}$$

and so on. It is these versions that are incorporated in the program.

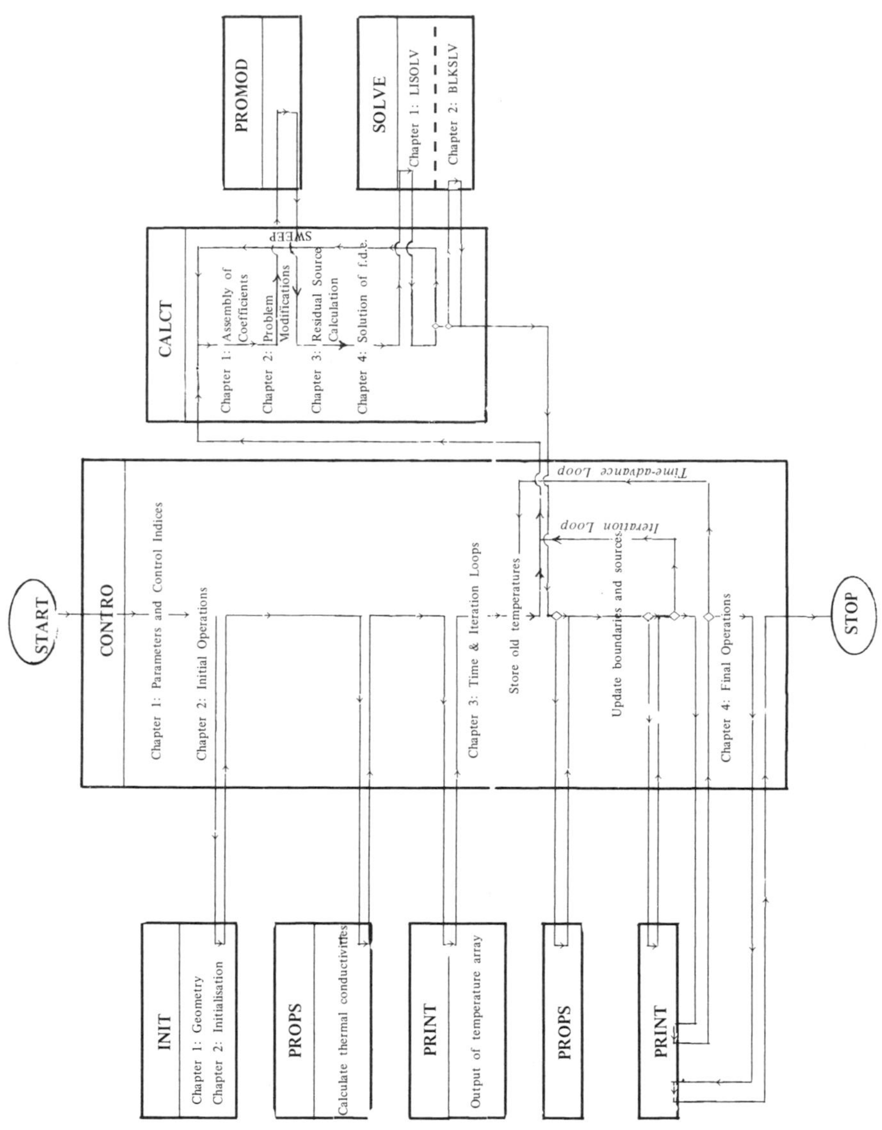

Fig. 0.7 Flow chart of the TEACH–C program

3.2 Overall structure of the program

By way of an introduction to TEACH–C, fig. 0.7 provides an overview, in the form of a flow chart showing the various subroutines, their functions and inter-relations, and the following paragraphs supplement the information given in the figure.

The core of TEACH–C is the main subprogram, here called 'CONTRO': this, as its name implies, has a *contro*lling function, which includes initialisation of the calculations according to user-supplied parameters, monitoring and controlling the progress of the calculations performed by the program as they proceed through the specified number of time steps, pausing to iterate at each step, and producing the resultant temperature fields and other information at intermediate and final stages. As indicated in the chart, CONTRO, like the majority of the subroutines, is subdivided, for ease of understanding, into 'Chapters' each of which is concerned with one of the major functions just outlined above.

The calculations proper are performed by a number of subsidiary subroutines comprising INIT, PROPS and CALCT, the last of which relies on the further subroutines PROMOD and SOLVE to perform its operations.

INIT, as the name implies, assists in the *init*ialisation process, its main function being to calculate the grid spacing and related quantities such as the dimensions of the control volumes.

PROPS is invoked to compute the material *prop*ertie*s* consisting in the present instance of the nodal values of the thermal conductivities. The option is available of calling PROPS repeatedly from within the iteration cycle, to allow for situations where k may vary.

The major *calc*ulations of *t*emperatures are performed by CALCT: each time this is called, a cycle of iteration is performed on the temperature field. CALCT operates by first assembling the coefficients of the fde (finite difference equation) at every interior cell, treating each in exactly the same way: it then calls on PROMOD to incorporate the *pro*blem *mod*ifications, consisting of the insertion of boundary conditions and sources, as appropriate. Finally SOLVE is called to *solve* the finite difference equations: this is effectively two subroutines, since it has two entry points named LISOLV, following which a sweep of the *li*ne *solv*ing procedure is applied and BLKSLV, which applies the *bl*oc*k* *solv*ing procedure.

The sole remaining subroutine is PRINT, whose function is to *print* out the temperature field, with appropriate headings.

TEACH–C has been deliberately structured so that the adaptation of the program to different situations, entails changes to (at most) three of the above subroutines, namely CONTRO, PROMOD and PROPS. The remaining subroutines are problem-independent, although provision has been made in them for easy exploration of different finite-difference formulations and/or solving procedures: thus the former will usually entail changes to CALCT only, while the latter will be confined to SOLVE.

Further details of the subroutines will be provided in section 3.5 below.

3.3 Important conventions and symbols

3.3.1 *Grid*

Fig. 0.8 shows the FORTRAN symbols denoting the coordinates of the grid nodes and various related quantities such as the inter-node spacings, control volume dimensions etc. The nodes are referenced by an (I, J) indexing system, with I denoting the grid line with x coordinate X(I) and J the line with y coordinate Y(J). The limits on I and J are $1 \leqslant I \leqslant NI$ and $1 \leqslant J \leqslant NJ$, but these do *not* define the domain of solution: separate arrays are provided for this purpose, denoted by JS(I) and JN(I), which confine the temperature calculations on grid line I to cells having J-values lying in the range $JS(I) \leqslant J \leqslant JN(I)$. This facility enhances computing economy for non-rectangular domains such as the example of fig. 0.2.

Selection of one of the three coordinate frames available (i.e. the Cartesian and two biradial frames) is made by way of the control parameters INCYLX and INCYLY: when both of these logical variables are specified as .FALSE., the Cartesian frame is selected and the program automatically sets the radii of curvature RX(I) and RY(J) to 1, the latter variables standing for r_x and r_y respectively. When either (but never both) INCYLX or INCYLY is set to .TRUE., the appropriate radius of curvature is calculated and stored in the relevant array, the other being set to unity.

Values for NI, NJ, X(I), Y(J), JS(I), JN(I), INCYLX and INCYLY are supplied by the user as input data. The program proceeds to calculate from these certain derived quantities shown in fig. 0.8. They include*:

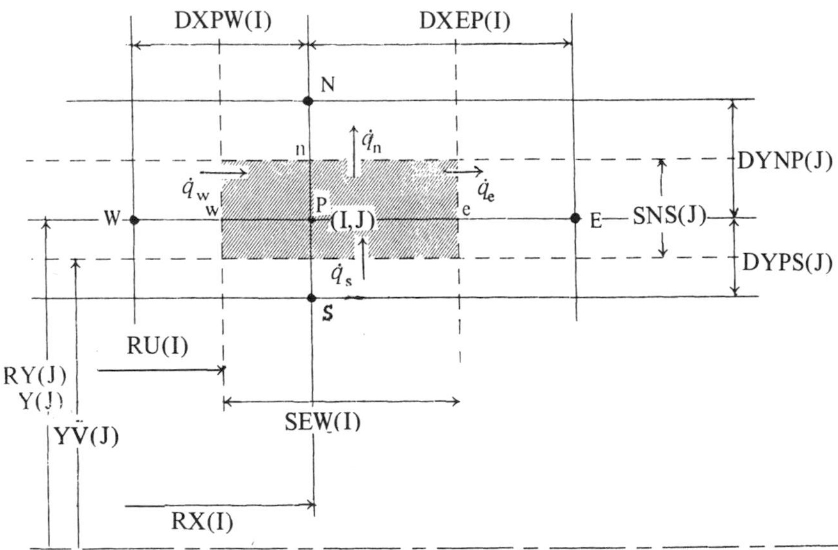

Fig. 0.8 FORTRAN notation for grid variables

* A full list of FORTRAN variables and their definitions is given in **Appendix A.**

DXPW(I), the inter-node spacing δx_{PW} between P and W;

DYPS(J), the inter-node spacing δy_{PS} between P and S;

SEW(I), the direction-x dimension δx_{ew} of the cell surrounding P;

SNS(J), the direction-y dimension δy_{ns} of the P cell.

Some of these arrays, it could be argued, are redundant: however they have been included to enhance understanding. In any case the additional storage requirements are negligible.

3.3.2 *Dependent variables and material properties*

The 'new' and 'old' temperatures are stored as T(I, J) and TOLD (I, J) respectively, where I and J denote the location, in the manner described above. An array is also provided for the thermal conductivities, named GAMH(I, J), so as to allow for spatial variations in this quantity. However, in conformity with the decision made in deriving the f.d.e. no such allowance is made for density and specific heat, the FORTRAN names of which are DENSIT and CV respectively.

The declared (I, J) dimensions of the above and other two-dimensional arrays in DIMENSION and COMMON statements must also be supplied as the variables IT and JT, for use in various locations in the program.

3.3.3 *Control parameters*

In addition to INCYLX and INCYLY, other control parameters exist which select important options in the program according to values assigned by the user. They are the following:

(*a*) *Specification of a time-dependent or steady-state calculation:–* this is performed through the logical variable INTIME which, when set equal to ·TRUE·, causes the calculations to proceed in a time-dependent fashion, as illustrated in fig. 0.7. When INTIME = ·FALSE·, the calculation of the time-dependent term $D_P(T_P^{new} - T_P^{old})$ of the f.d.e. is bypassed and the program simply iterates to the final solution. This practice is equivalent to, but more economical than, setting the time increment δt to an effectively infinite value.

(*b*) *Selection of constant or variable thermal conductivity:–* if k is expected to vary throughout the calculations as a consequence of, e.g., temperature dependence, setting the variable INPRO = ·TRUE· causes PROPS to be called repeatedly from within the iteration cycle and therefore allows k to be updated according to a user-specified formula. INPRO = ·FALSE· suppresses this option.

(*c*) *Control of calculations.* The solution cycle is controlled through the following parameters:

MAXSTP	is the ***max***imum number of time ***step***s to be performed (when INTIME = ·FALSE·, MAXSTP is automatically set equal to 1). A running count of the ***n***umber of time ***step***s performed is stored as NSTEP.
MAXIT	is the ***max***imum allowable ***it***eration cycles at each time step (if this limit is reached an appropriate message is printed out and the calculations are terminated). The running count of the ***n***umber of ***iter***ations is NITER.
SORMAX	is the level of the normalised residual source sum $\sum_{i,j} \|R_P\|$ at which the iterative procedure is taken to be converged and the program exits from the iteration loop. The (sum of the absolute values of) the un-normalised sources (see eqn. (0.27)) is RESORT and the normalising factor is SNORM.
DT	is the specified time increment δt which may be varied by the user during the course of the calculations.
URFT	is the ***u***nder-***r***elaxation ***f***actor on ***t***emperature, defined in eqn. (0.31).

(*d*) *Control of output.* Provision is made in TEACH–C for the output of information of various kinds during the course of the calculations. Some of this information is governed by user-specified control parameters, as follows:

NITPRI	designates the interval, measured as a ***n***umber of ***it***erations, at which the temperature field is ***pri***nted out during the iteration process.
NSTPRI	is the interval, measured in ***n***umber of time ***st***eps, at which the converged solution is ***pri***nted out.
IMON, JMON	are the (I, J) indices of a point in the field at which the temperature is produced at every iteration, for ***mon***itoring purposes.

3.3.4 *Coefficients of finite-difference equations*

The FORTRAN equivalents of the various coefficients appearing in the finite-difference equation (0.3) are:

Coefficient	*FORTRAN symbol*
A_N	AN (J)
A_S	AS (J)
A_E	AE (J)
A_W	AW (J)
$A_P - S_P$	AP (J)
S_U	SU (J)
S_P	SP (J)

Here, it should be noted, only *one*-dimensional arrays are used for storage, these being overwritten as the calculations proceed from one grid line to the next.

3.4 Description of the problem solved by the basic version of TEACH–C

3.4.1 *Introductory remarks*

The basic version of TEACH–C listed in Appendix C has been set up for illustrative purposes to calculate a particular transient two-dimensional conduction problem. A description of this problem will therefore be presented before the details of the individual subroutines are given, so that these may be seen in context.

3.4.2 *The nature of the problem*

Fig. 0.9 illustrates the cross-section of a long rectangular bar, initially at zero temperature everywhere. A constant temperature $T = 100$(these values are arbitrary) is suddenly imposed at the upper surface, while the remaining surfaces are held at $T = 0$. The task is to compute the development of the temperature field within the bar to the final steady-state condition.

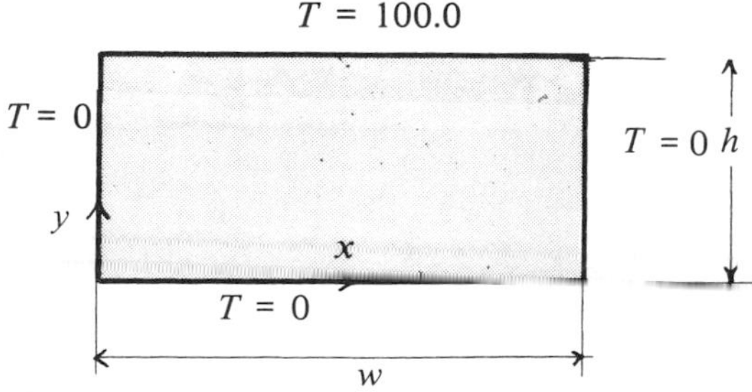

Fig. 0.9 The problem solved by the basic version of TEACH–C

3.4.3 *Relevant FORTRAN variables*

The following variables in the basic TEACH–C program are particular to the demonstration problem:

H	the height h of the bar;
W	the width w of the bar;
TBOT TTOP TLEFT TRIGHT	the temperatures of the bottom, top and sides, as viewed in fig. 0.9

3.5 Description of the individual subroutines

3.5.1 *Introduction*

In this section further details are provided of the individual subroutines of TEACH–C: the material contained herein is intended to be read in conjunction with the listing of the program appearing in Appendix C. The problem-dependent subroutines are described first, and then the remaining subroutines are presented.

3.5.2 *Problem-dependent subroutines*

(*a*) *CONTRO*

The general role of this subroutine has been explained earlier in this Chapter. Accordingly, attention will here be focussed on features which have not yet received mention, particularly those relating to the specification of the illustrative problem.

(*i*) *Chapter 0* is concerned with preliminary operations, of which the more noteworthy ones are: the specification of alphanumeric data for output headings (stored in the arrays HEDT (6) and HEDS (6)), and the declaration of the array dimensions IT and JT.

(*ii*) *Chapter 1* commences with the grid, Cartesian co-ordinates are specified by setting both INCYLX and INCYLY equal to ·FALSE· and the values assigned to NI and NJ designate a 12 x 12 mesh (so there are 10 cells in each direction, since the external nodes are inactive). The data are for a square bar of side unity (i.e. H = W = 1) and although provision is made here for calculating a regularly-expanding grid*; specification of the expansion factors FEXPX and FEXPY as unity yields, in this instance, uniform spacing for the coordinates X(I) and Y(J) of the nodes. (It should be recalled that the convention is to arrange the grid such that the problem boundaries coincide with cell boundaries: here the specification locates the former at $x = w, 0$ and $y = 0, h$). Finally, the traverse limits JS(I) and JN (I) are set at 2 and (NJ – 1) respectively.

The material properties specified are those of steel, in S.I. units.

* According to X(I) – X(I–1) = FEXPX*[X(I–1) – X(I–2)] and similarly for Y(J).

The values assigned to the control parameters specify 20 time steps of 50s duration, with a maximum of 100 iterations allowed at each step. The temperature array is to be printed out at intervals of 20 iterations (NITPRI = 20) and at each time step (NSTPRI = 1). No under-relaxation is specified (URFT = 1.0) and the criterion for convergence is SORMAX = 10^{-4}. The monitoring location for temperature printout during iteration is set at IMON = 6, JMON = 6.

(*iii*) *Chapter 2* commences with a call to INIT, which has the effect of causing the geometrical parameters of the grid to be calculated and the arrays to be initialised with zero values. Then, back in CONTRO, sequences appear for assigning the boundary temperatures (here it should be noted, the redundant external grid locations are employed for storing the boundary values) and inserting the initial field of internal temperatures: the call to PROPS then initialises the thermal conductivity array. The next operation is to calculate the residual source normalisation factor SNORM*, here obtained from a notional thermal resistance formula:

$$k\,(T_{top} - T_{bot})\,w/h$$

The final operation in this chapter is the printing out of problem-specification information according to instructions supplied by the user.

(*iv*) *Chapter 3* which is concerned with the main calculations, has required no alteration for the present application. In general however changes may be required, of the following kinds:

- ♦ running adjustments to DT during unsteady calculations, so as to minimise computing times.
- ♦ updating of boundary conditions and/or heat sources which are time- or temperature-dependent.
- ♦ changes to the quantities printed out as monitoring information.

The appropriate locations for these are indicated by COMMENT cards. One of the more important functions of this chapter is the monitoring of the progress of the iterative procedure: of particular note is the check made on the normalised residual source at NITER$\geqslant$20, which causes the program to halt if the source exceeds a prescribed value, thus indicating divergence. An appropriate message is printed out in this case.

(*v*) *Chapter 4* where the final operations are performed following termination of the main calculations, here simply provides an output of the final temperature field. Additional operations may, of course, be carried out according to the requirements of the user.

(*b*) *PROMOD*

In this subroutine, the problem-modification subroutine sequences are provided for incorporating the boundary conditions of the f.d.e.'s according to the general procedure outlined in Section 2.2(*c*). Since in the present example all four boundaries have prescribed temperatures, it will suffice to examine the treatment of just one of them, say the 'North' wall. Inspection of the relevant sequence will reveal that:

* It is important to note that SNORM is a problem-dependent quantity which *must be specified by the user*. The specification shown is appropriate only to the demonstration case.

(*i*) the first operation is to suppress the inappropriate heat-flow component in the f.d.e. by setting AN(NJ–1) = 0;

(*ii*) then the correct heat flow expression, which in algebraic form runs:

$$\dot{q}_n = k_B \frac{(T_B - T_P)\,\delta x_{ew}}{\delta y_{BP}} \tag{0.36}$$

where the notation is that of Fig. 0.4, is inserted via SU and SP, the quantity $k_n\,\delta x_{ew}/\delta y_{nP}$ being represented by DN.

The remaining boundaries are similarly treated; an important point to note is that the alterations to SU and SP are made in an *additive* fashion so as not to interfere with already-existing contents. The latter might, for example, consist of distributed heat sources, which would also be inserted in PROMOD.

It should also be noted that use is made of the index IL, which contains the *I* value of the grid *l*ine being solved, to determine when to apply the modifications for the side boundaries.

(*c*) *PROPS*

This subroutine requires attention only when the thermal conductivity varies in time or space in some manner which must be specified here by the user. In the present example the thermal conductivity is constant, so the specified value TCON is simply fed into the GAMH (I, J) array in a once-for-all operation.

3.5.3 *Problem-independent subroutines*

(*a*) *CALCT*

(*i*) *Chapter 1* of this subroutine is concerned with the assembly of the f.d.e. coefficients. This is done in a line-by-line fashion for $2 \leqslant I \leqslant NI-1$ the scanning limits along each line being determined from JS(I) and JN(I). Advantage is taken in the coefficient assembly of the fact that they obey reciprocity relations, i.e. AS (J) = AN (J–1) and AW (J) for line I is equal to AE (J) for line I–1: thus only two coefficients are calculated for each cell.

(*ii*) *Chapter 2* simply contains the call to PROMOD which causes the coefficients to be modified where necessary.

(*iii*) *Chapter 3* contains instructions for: the assembly of the difference equations into the form required for solution; the calculation and summing of the residual sources*; and the incorporation of under-relaxation, according to eqn. (0.32).

(*iv*) *Chapter 4* invokes the solving procedures, through calls to LISOLV and BLKSLV.

* These are computed, for reasons of economy, from the prevailing temperature field, which is made up of values from the current and previous iteration. This should be borne in mind when interpreting the behaviour of the sources.

(*b*) *SOLVE*

(*i*) *Chapter 1,* with entry point LISOLV, contains the TDMA procedure (see Appendix B). The sequence is first to assemble the coefficients in eqn. (0.25), then calculate the coefficients of the recurrence relation (0.26) and finally solve for the temperatures by back substitution. Also embedded in this sequence is the assembly of the coefficients of eqn. (0.29) required for the block adjustment procedure (see below).

(*ii*) *Chapter 2,* with entry point BLKSLV, invokes the block adjustment procedure of Section 2.3. Because the coefficients for this have already been assembled in Chapter 1, all that is done here is to obtain the solution using the TDMA, and then make the adjustments.

Users who wish to explore alternative solving procedures may easily do so by replacing SOLVE by their version and suitably altering CALL statements in CALCT.

(*c*) *INIT*

(*i*) *Chapter 1* contains the grid calculations, which are self-explanatory It should be noted that the coordinates of the outermost grid lines are adjusted to coincide with those of the outermost cell boundaries, so as to facilitate interpretation of the results.

(*ii*) *Chapter 2* is the initialisation section.

(*d*) *PRINT*

This is a general-purpose output subroutine which produces the contents of any two-dimensional array PHI (I,J) as an ordinary array of numbers with a title specified in the alphanumeric array * HEAD, both of the fore-mentioned variables being arguments of the subroutine. Usually PHI(I,J) is replaced in the calling statements by T(I,J) and HEAD by HEDT, the latter containing the name 'TEMPERATURE'; then the output is as shown in Appendix D. However, the user may also employ PRINT for example to print out the thermal conductivities, with an appropriately chosen heading, by suitable alteration of the calling statement. (It should be noted that a spare alphanumeric array HEDS is available for this purpose.)

* Dimensioned 6 (A6), which allows a 36-character string. This specification will require changing for machines with word lengths accommodating fewer than 6 characters.

3.6 Suggestions for preparation and operation of the program

3.6.1 *Preparation*

The following 'check-list' is intended as a guide for users who wish to adapt TEACH–C to their particular requirements:

(*i*) *In CONTRO:*

(*a*) Specify in Chapter 0 the dimensions (IT, JT) of the two-dimensional arrays, which should correspond to the declared dimensions in the common statements.

(*b*) *In Chapter 1:*

- Select the coordinate frame via INCYLX and INCYLY; declare the number of grid intervals through NI and NJ; provide sequences for calculating the grid coordinates X (I) and Y (I), bearing in mind the convention that the physical boundaries are located mid-way between grid-lines, and set the upper and lower traverse limits JN (I) and JS (I) for each line.
- Assign appropriate values to the control parameters INTIME and INPRO. If the former specifies an unsteady calculation, assign values to the time increment DT and the output interval NSTPRI: for steady-state problems these may be ignored. If INPRO specifies temperature-dependent properties, insert in PROPS the appropriate sequences for calculating these.
- Provide values for the material properties TCON, CV and DENSIT.
- Specify the iteration-control parameters MAXIT, URFT and SORMAX.
- Specify the printout-control quantities NITPRI, NSTPRI and IMON, JMON.

(*c*) *In Chapter 2:*

- Provide sequences for inserting fixed boundary temperatures, where applicable (recall that these are stored, for convenience, at external grid locations, but of course the physical location of these nodes is without significance) and initial values within the field.
- Arrange for the calculation of an appropriate source-normalisation factor SNORM.
- Arrange for print-out of problem-specification information.

(*d*) *In Chapter 3:*

- Insert, when necessary, instructions for updating boundary temperatures and heat sources.
- If special output is required at each time step, insert the necessary instructions just below statement 3500

(*e*) Provide in Chapter 4 sequences for the processing and print-out of the final results.

(*ii*) *In PROMOD:* provide instructions for the insertion of boundary conditions and source terms appropriate to the problem, bearing in mind that

- ♦ conditions must be specified at *all* boundaries;
- ♦ the procedure for inserting the boundary conditions is to break the normal link, and then insert the correct one via the linearised-source coefficients;
- ♦ 'true' heat sources should also be linearised where possible and inserted via the source coefficients;
- ♦ the alterations to the source coefficients should be made in an additive fashion, so as not to interfere with earlier ones.

3.6.2 *Operation*

The simplicity of the tasks which TEACH–C performs, and of the program itself, renders it comparatively easy to use. However, experience has suggests that the following additional advice may be helpful:

(*a*) *Specification of calculation parameters*

The most general class of problems which TEACH–C is capable of solving are unsteady, two-dimensional and non-linear, due to temperature-dependent thermal conductivity or heat sources. For these, it is necessary to march out the solution through time, pausing at each time step to iterate until convergence is achieved and recalculating the finite-difference coefficients at each iteration as would normally be done.

If, however, not all the complexities metioned are present in the situation considered, some optimisation is possible by appropriate specification of the control parameters, as is illustrated by the following examples:

(*i*) For *steady-state problems* it is always more economical of computing time to proceed directly to the solution via iteration on the steady-state equations, by setting INTIME = ·FALSE· (setting DT to a large, effectively infinite, number has a similar effect, but requires slightly more computing time due to the (redundant) calculation of the unsteady terms).

(*ii*) For *one-dimensional problems* it is advantageous to arrange that the calculation is in the direction of the line-iteration procedure, i.e. in the y or N–S direction. This will ensure maximum economy, for it has been arranged that, when the grid is appropriately reduced (i.e. NI is set to 3) the block-adjustment procedure, which is now rêdundant, is bypassed. Indeed, for *linear* one-dimensional problems just two 'passes' through CALCT will be required, the second one being needed merely to confirm that the residual sources have been reduced to zero.

(*iii*) For *two-dimensional problems* it is preferable to arrange for the block adjustments to be made along grid lines running perpendicular to the direction of predominant temperature gradient, for this will minimise the number of iterations required.

For those users restricted more in respect of computing time than in computer storage, further reductions in the former can be achieved by arranging for the f.d.e. coefficients to be stored in *two-dimensional* arrays (i.e. AN (J) is replaced by AN (I, J)), assembled over the *whole* field before SOLVE is invoked and then for the latter to be called repeatedly, using the same coefficients each time. This arrangement was successfully employed in an earlier version of TEACH–C, although it was advantageous only for the special class of problems where the coefficients were temperature-independent, or nearly so.

(*b*) *Checks for convergence and accuracy*

It is important, before accepting a solution as representing the physical problem being modelled, to make the following checks.

(*i*) Ensure that the iterative procedure has converged, i.e. that SORMAX and SNORM have been specified such that, at each time step the residual sources have been reduced to acceptably small values and the temperature field changes insignificantly over the last few iterations.

(*ii*) Obtain solutions with successively smaller grid and time intervals, until further reduction produces no appreciable changes.

(*c*) *Suppression of numerical instability*

Apart from errors in formulation or programming, the main sources of numerical instability or 'divergence' (which will be manifested by growing or oscillating residual sources) are non-linearities. If the latter are the result of temperature-dependent thermal conductivities, under-relaxation (achieved by assigning to URFT a value less than unity) will normally procure convergence.* The same cure will often suffice for non-linear heat sources or boundary conditions, but here it is useful also to ensure that these are linearised when introduced and that the linearisation is such as to ensure that the coefficient $\boldsymbol{B}_P$ of the linearised-source expression is always negative.

---

* Strictly, of course, all references to the achievement of convergence should be read as *achieving satisfaction of the criterion selected for convergence.* No finite numerical procedure of the kind chosen can ever determine convergence in the strict mathematical sense.

4. Closing Remarks

As we remarked in the Introductory section, there are many physical phenomena which can be simulated with the aid of the TEACH–C program and there are equally many ways of using such computer explorations as instructional tools. We hope that the material in this Guide together with the PROBLEMs that follow will suggest to the reader how to use the program in the way best suited to his or her purposes.

We gratefully acknowledge the support of the National Development Programme in Computer-Assisted Learning as well as the assistance and encouragement of a number of staff and students in the Mechanical Engineering Department, Imperial College, London.

5. References

0.1 A.D. Gosman and F.J.K. Ideriah. *The TEACH–2E program for calculating turbulent recirculating flows in two dimensions.* Report in preparation.

0.2 A.D. Gosman and S.A. Syed. *The TEACH–3P program for calculating three-dimensional boundary-layer flows.* Report in preparation.

0.3 A.D. Gosman, B.E. Launder, F.C. Lockwood, G.J. Reece, *A C.A.L. Course in Fluid Mechanics and Heat transfer,* presented at Computers in Higher Education Conference, Loughborough, March 1976. Reprinted in *International Journal of Mathematical Education in Science & Technology,* Volume 8, No. 1, 1–15, 1977.

0.4 A.D. Gosman, B.E. Launder, F.C. Lockwood, G.J. Reece, *Computer Assisted teaching of Fluid Mechanics and Heat Transfer,* in "Proceedings of the Seventh Conference on Computers in the Undergraduate Curricular", Binghampton, New York, June 1976. Reprinted in *Computers in Education,* Volume 1, No. 3, 1977.

0.5 A.D. Gosman, B.E. Launder, F.C. Lockwood, P.A. Newton, G.J. Reece, *Applications of TEACH–C: PROBLEM 1 Unsteady One-dimensional Conduction Processes,* Imperial College, Mechanical Engineering Department Report CAL–C1–76.

0.6 – PROBLEM 2 *Two dimensional Conduction Processes,* CAL–C2–76

0.7 – PROBLEM 3 *Irrotational Stagnation Flow,* CAL–C3–77

0.8 – PROBLEM 4 *Fully-developed Flow in Ducts,* CAL–C4–77

0.9 – PROBLEM 5 *Heat Transfer in Fully-developed Flow in Non-Axisymmetric Ducts,* CAL–C5 (in preparation)

0.10 – PROBLEM 6 *Convection in Developing Duct Flows,* CAL–C6 (in preparation)

0.11 – PROBLEM 7 *Convection in Stagnation Flow,* CAL–C7 (in preparation)

0.12 – PROBLEM 8 *Three-dimensional Convection in Developing Duct Flow,* CAL–C8 (in preparation)

0.13 P.J. Schneider, *Conduction Heat Transfer,* Addison-Wesley, 1955.

0.14 E.R.G. Eckert, *Introduction to the transfer of Heat and Mass,* McGraw-Hill, 1950.

0.15 R.S. Varga, *Matrix Iterative Analysis,* Prentice-Hall, 1962.

0.16 G.D. Smith, *Numerical Solution of Partial Differential Equations,* Oxford University Press, 1965.

0.17 J.R. Westlake, *A Handbook of Numerical Matrix Inversion and Solution of Linear Equations,* John Wiley, 1968.

6. Notation

Symbol	*Meaning*
A_E, A_N, A_S, A_W	coefficients in finite-difference equation (0.23) representing thermal conductances.
a_j, b_j, c_j, d_j	coefficients in TDMA equation (0.25)
a'_j, c'_j	coefficients in recurrence relation (0.26)
A_P	$\Sigma_c A_c$, coefficient in finite-difference equation (0.23)
B_P	coefficient in linearised-source equation (0.15)
C_P	coefficient in linearised-source equation (0.15)
c_v	constant-volume specific heat capacity
D_P	$\rho_P c_v V_P / \delta t$, coefficient in finite-difference equation (0.23) representing thermal capacitance of cell
e, n, s, w	labels of control-volume boundaries in fig. 0.3
E, N, P, S, W	Labels of nodes in typical grid 'molecule' (see fig. 0.3)
f_1, f_2, f_3	coefficients in general boundary condition (eqn. (0.5))
k	thermal conductivity
n	coordinate normal to a boundary
δn	increment of distance along normal to boundary
r	radial coordinate in a cylindrical-polar frame
s_P	$-(D_P - B_P)$, coefficient in substitution formula (0.23
$\delta r_{NP}, \delta r_{PS}$	radial grid spacings (see fig. 0.3)
δr_{ns}	radial dimension of cell
s_T	heat-generation rate per unit volume of material
t	time
T	temperature
δt	time increment
V_P	$r_P \delta r_s \delta x_{ew}$, approximate volume of cell
x	Cartesian coordinate
y	Cartesian coordinate
$\delta x_{EP}, \delta x_{Pw}$	axial grid spacings (see fig. 0.3)
δx_{ew}	axial dimension of cell

Greek Characters

β	under-relaxation factor, defined in eqn. (0.31)
γ	'large number' used in temperature-fixing procedure
ρ	density

Subscripts

B	pertaining to physical boundary
E, N, P, S, W	pertaining to grid nodes of typical cluster in fig. 0.4
e, n, s, w	pertaining to control-volume boundaries
N	denotes Nth iteration
r, x	pertaining to coordinate directions r, x

Superscripts

old	denotes old value
new	denotes new value

Appendix A: Glossary of FORTRAN Notation

FORTRAN Name	*Meaning*
A (J)	coefficient A_j in TDMA equation
AE (J)	coefficient A_E in finite-difference equation
AN (J)	coefficient A_N in finite-difference equation
AP (J)	coefficient $A_P - S_P$ in finite-difference equation
AS (J)	coefficient A_S in finite-difference equation
AW (J)	coefficient A_W in finite-difference equation
B (J)	coefficient b_j in TDMA equation
C (J)	coefficient c_j in TDMA equation
CV	specific heat at constant volume
D (J)	coefficient D_j in TDMA equation
DENSIT	density ρ of the material
DT	time increment δt
DXEP (I)	inter-node spacing δx_{EP} = X(I + 1) – X(I)
DXPW (I)	inter-node spacing δx_{PW} = X(I) – X(I – 1)
DYNP (J)	inter-node spacing δy_{NP} = Y(J + 1) – Y(J)
DYPS (J)	inter-node spacing δy_{PS} = Y(J) – Y(J – 1)
GAMH (I, J)	the local thermal conductivity k at node (I, J)
GREAT	the 'large number' γ employed in the temperature-fixing procedure
H	the height h of the rectangular bar
HEDS (6)	array containing alphanumeric data for headings
HEDT (6)	array containing alphanumeric data for headings
I	index denoting axial location in grid
IMON	I index of a point in the field at which the temperature is produced at every iteration for monitoring purposes
INCYLX	logical variable to select (INCYLX = .TRUE.) cylindrical polar frame with radial coordinate in x direction.
INCYLY	logical variable to select (INCYLY = .TRUE.) cylindrical polar frame with radial coordinate in y direction.

FORTRAN Name	*Meaning*
INPRO	logical variable which activates or suppresses recalculation of thermal conductivities according to convention: ·TRUE· = activate, ·FALSE· = suppress
INTIME	logical variable which selects time-dependent or direct path to steady-state solution according to convention ·TRUE· = time-dependent, ·FALSE· = direct
IT	the declared dimension of I in all two-dimensional arrays
J	index denoting radial location in grid
JMON	the J index of a point in the field at which the temperature is produced at every iteration for monitoring purposes
JN (I)	array containing the cell numbers of the upper limit of the calculation domain for each vertical line of the grid
JS (I)	array containing the cell numbers of the lower limit of the calculation domain for each vertical line of the grid
JT	the declared dimensions of J in all two-dimensional arrays
MAXIT	maximum allowable number of iterations at each time step
MAXSTP	total number of time steps to be performed
NI	the total working number of vertical grid lines
NIM1	$\equiv$ NI $-$ 1
NITER	a running counter of the number of iteration performed
NITPRI	the interval, measured in number of iterations, at which the temperature field is printed out during the iteration process
NJ	the total working number of horizontal grid lines
NJM1	= NJ $-$ 1
NSTEP	a running counter of the number of time steps performed
NSTPRI	the interval, measured in number of time steps, at which the converged solution is printed out

FORTRAN Name	*Meaning*
PI	π (= 3.14159265...)
RESORT	the sum of the absolute values of the residual energy sources
RU (I)	the radial distance from the symmetry axis to the west boundary of the cell (I, J) when INCYLX = ·TRUE·
RV (J)	the radial distance from the symmetry axis to the south boundary of the cell (I, J) when INCYLY = ·TRUE·
RX (I)	the radial distance of node (I, J) from the symmetry axis when INCYLX = ·TRUE·
RY (J)	the radial distance of node (I, J) from the symmetry axis when INCYLY = ·TRUE·
SEW (I)	the axial dimension of the cell (I, J) (see fig. 0.8, page 26)
SNORM	a characteristic heat flow rate employed to normalise the total residual energy source
SNS (J)	the radial dimension of the cell (I, J) (see fig. 0.8, page 26)
SOURCE	the normalised total residual source
SORMAX	the level of the normalised total residual source at which the calculations are taken as converged
SP (J)	the coefficient S_P in the finite-difference equation
SU (J)	the coefficient S_U in the finite-difference equation
T (I, J)	temperatures
TBOT	the temperature of the bottom of the rectangular bar
TCON	the reference value of the thermal conductivity
TIME	elapsed time covered by calculations
TLEFT	temperature of the left side of rectangular bar
TOLD (I, J)	'old' temperatures
TRITE	temperature at right side of rectangular bar
TTOP	temperature at the top of rectangular bar
URFT	under-relaxation factor β
W	the width w of the rectangular bar
X (I)	horizontal coordinate of node (I, J)

XU (I)	the horizontal location of the west boundary of cell (I, J)
Y (J)	the vertical coordinate of node (I, J)
YV (J)	the vertical coordinate of the south boundary of cell (I, J)

APPENDIX B

The Tri-Diagonal Matrix Algorithm (TDMA)

The set of equations to be solved for each grid line has the form

$$d_j T_j = a_j T_{j+1} + b_j T_{j-1} + c_j$$

which is eqn. (0.25) of the main text.

If there are n cells along a particular line, then j will vary from 2 to $J = n-1$. More fully, the equations are:

$$\begin{aligned} d_2 T_2 &= a_2 T_3 + c_2 \\ d_3 T_3 &= a_3 T_4 + b_3 T_2 + c_3 \\ d_4 T_4 &= a_4 T_5 + b_4 T_3 + c_4 \\ &\vdots \\ d_{n-2} T_{n-2} &= a_{n-2} T_{n-1} + b_{n-2} T_{n-3} + c_{n-2} \\ d_J T_J &= + B_J T_{n-2} + c_{n-1} \end{aligned} \tag{B.1}$$

It should be noted that b_2 and a_{n-1} are always zero as a consequence of the practice of setting the coefficients A_i relating to points external to boundaries to zero, and incorporating the boundary conditions through the source coefficients S_v and S_p, which are contained in

$$d_2,\ c_2 \text{ and } d_{n-1},\ c_{n-1}.$$

The set of equations (B.1) is solved by a forward-elimination procedure followed by a back-substitution procedure. The first of these procedures eliminates the T_{j-1} term from each equation by substitution from the previous equation as follows: we write the first of eqns. (B.1) as

$$T_2 = a_2' T_3 + c_2'$$

where $a_2' = a_2/d_2$ and $c_2' = c_2/d_2$. Substituting this into the second equation gives

$$(d_3 - b_3 a_2') T_3 = a_3 T_4 + c_3 + b_3 c_2',$$

or

$$T_3 = a_3' T_4 + c_3'$$

where

$$a_3' = \frac{a_3}{(d_3 - b_3 a_2')} \text{ and } c_3' = \frac{(c_3 + b_3 c_2')}{d_3 - b_3 a_2'}$$

This result may now be substituted into the third equation to give:

$$T_4 = a_4' T_5 + c_4'$$

where

$$a_4' = \frac{a_4}{(d_4 - b_4 a_3')} \text{ and } c_4' = \frac{(c_4 + b_4 c_3')}{(d_4 - b_4 a_3')}$$

and so on.

When this elimination is performed for all equations (B.1) there results the new set:

$$\begin{aligned} T_2 &= a_2' T_3 + c_2' \\ T_3 &= a_3' T_3 + c_3' \\ &\vdots \\ T_{n-2} &= a'_{n-2} T_{n-1} + c'_{n-2} \\ T_J &= c'_{n-1} \end{aligned} \qquad \text{(B.2)}$$

which can be solved by the application of the back-substitution procedure. The last of eqns. (B.2) gives a value for T_J directly: this can be substituted into the penultimate equation to give a value for T_{n-2}; and so on until the final value T_2 is obtained.

APPENDIX C – Listing of TEACH–C

```
      PROGRAM CONTRO(INPUT,OUTPUT,TAPE5=INPUT,TAPE6=OUTPUT)
C***********************************************************************
C
C                                TEACH-C
C
C        A COMPUTER PROGRAM FOR THE SIMULATION OF HEAT CONDUCTION,
C                   CONVECTION AND ANALOGOUS PHENOMENA
C
C***********************************************************************
C
C      * DEVELOPMENT OF THIS PROGRAM HAS BEEN SUPPORTED BY THE
C        NATIONAL DEVELOPMENT PROGRAMME IN COMPUTER ASSISTED LEARNING
C        THROUGH GRANT DP/102A.
C
C
C
C
```



```
C
C
C
C
C **********************************************************************
C      SUBROUTINE CONTRO
C-----------------------PROGRAM-CONTROL ROUTINE--------------------------
C
CHAPTER  0  0  0  0  0  0  0  0  PRELIMINARIES    0  0  0  0  0  0  0  0
C
C      COMMON
C     1/CONVAR/IT,JT,INTIME,DT,RESORT,UREF,GREAT
     1/PROP/TCON,CV,DENSIT,GAMH(22,22)
     1/TEMP/T(22,22),TOLD(22,22)
     1/GEOM/NI,NJ,NIM1,NJM1,INCYLX,INCYLY,DX,DY,RX(22),RU(22),
     1      X(22),Y(22),DXEP(22),DXPW(22),DYNP(22),DYPS(22),
     1      SNS(22),SEW(22),XU(22),YV(22),RY(22),RV(22),JS(22),JN(22)
     1/COEF/IL,AP(22),AN(22),AS(22),AE(22),AW(22),SU(22),SP(22)
      LOGICAL INPRO,INTIME,INCYLX,INCYLY
      CHARACTER*6 HEDT(6),HEDS(6)
      DATA HEDT(1),HEDT(2),HEDT(3),HEDT(4),HEDT(5),HEDT(6)
     1    /'      ','TEMPER','ATURE ','(C)   ',2*'      '/
      DATA HEDS(1),HEDS(2),HEDS(3),HEDS(4),HEDS(5),HEDS(6)
     1    /6*'      '/
      OPEN(5,FILE='INPUT')
      OPEN(6,FILE='OUTPUT')
      IT=22
      JT=22
      GREAT=1.0E30
      PI=4.*ATAN(1.0)
C
CHAPTER  1  1  1  1  1  PARAMETERS AND CONTROL INDICES  1  1  1  1  1  1
C
C-----SPECIFY GRID
C-----SET CARTESIAN/CYLINDRICAL COORDS. BY INCYLX/INCYLY=.FALSE./.TRUE.
      INCYLX=.FALSE.
      INCYLY=.FALSE.
      NI=12
      NJ=12
      W=1.0
      H=1.0
      FEXPX=1.0
      FEXPY=1.0
C-----CALCULATE X COORDINATES WITH EXPANSION FACTOR FEXPX
      DX=W/FLOAT(NI-2)
      IF (FEXPX.NE.1.0) DX=2.0*W*(1.0-FEXPX)
     1                   /(1.0+FEXPX-FEXPX**(NI-2)-FEXPX**(NI-1))
      X(1)=-0.5*DX
      DO 1100 I=2,NI
      X(I)=X(I-1)+DX
 1100 DX=FEXPX*DX
C-----CALCULATE Y COORDINATES WITH EXPANSION FACTOR FEXPY
      DY=H/FLOAT(NJ-2)
      IF (FEXPY.NE.1.0) DY=2.0*H*(1.0-FEXPY)
     1                   /(1.0+FEXPY-FEXPY**(NJ-2)-FEXPY**(NJ-1))
      Y(1)=-0.5*DY
      DO 1200 J=2,NJ
      Y(J)=Y(J-1)+DY
 1200 DY=FEXPY*DY
```

```
C-----SET VERTICAL TRAVERSE LIMITS
      DO 1300 I=1,NI
      JS(I)=2
 1300 JN(I)=NJ-1
C-----SET MONITORING LOCATION
      IMON=6
      JMON=6
C-----MATERIAL PROPERTIES
      TCON=52.0
      CV=460.0
      DENSIT=7850.0
C-----PROGRAM CONTROL PARAMETERS
      MAXIT=100
      MAXSTP=20
      NITPRI=20
      NSTPRI=1
C-----UNDER-RELAXATION FACTOR,MAXIMUM RESIDUAL SOURCE AND TIME INCREMENT
      URFT=1.0
      SORMAX=0.0001
      DT=50.0
C-----SELECTION OF STEADY (INTIME=.FALSE.) OR UNSTEADY OPTION
      INTIME=.TRUE.
      IF (.NOT.INTIME) MAXSTP=1
C-----SELECTION OF CONSTANT (INPRO=.FALSE.) OR VARIABLE-PROPERTY OPTION
      INPRO=.FALSE.
C
CHAPTER  2  2  2  2  2  2  INITIAL OPERATIONS  2  2  2  2  2  2  2  2  2
C
C-----CALCULATE GRID QUANTITIES AND SET ARRAYS TO ZERO
      CALL INIT
      TIME=0.0
C-----ASSIGN BOUNDARY VALUES AND INITIALISE INTERIOR FIELD
      TTOP=100.0
      TBOT=0.0
      TLEFT=0.0
      TRIGHT=0.0
      DO 2100 I=2,NIM1
      T(I,NJ)=TTOP
 2100 T(I,1)=TBOT
      DO 2200 J=2,NJM1
      T(1,J)=TLEFT
 2200 T(NI,J)=TRIGHT
C-----INITIALISE MATERIAL PROPERTY FIELD
      CALL PROPS
C-----CALCULATE RESIDUAL-SOURCE NORMALISATION FACTOR
C-----THIS CASE USES A NOTIONAL HEAT FLOW DEDUCED FROM
C-----A THERMAL RESISTANCE FORMULA
      SNORM=TCON*(TTOP-TBOT)*W/H
      SNORM=ABS(SNORM)
C-----PROVIDE OUTPUT OF PROBLEM SPECIFICATION
      WRITE(6,2900)H,W,CV,TCON,DENSIT,DT,SNORM,NI,NJ
      CALL PRINT(1,1,NI,NJ,IT,JT,X,Y,T,HEDT)
C
CHAPTER  3  3  3  3  3  3  TIME AND ITERATION LOOPS 3  3  3  3  3  3
C
      WRITE(6,3000)IMON,JMON
C-------------------------START OF TIME ADVANCE LOOP--------------------
      DO 3600 NSTEP=1,MAXSTP
      IF(.NOT.INTIME)GO TO 3200
      TIME=TIME+DT
      DO 3100 I=1,NI
      DO 3100 J=1,NJ
 3100 TOLD(I,J)=T(I,J)
C------------------------ START OF ITERATION LOOP -----------------------
 3200 DO 3400 NITER=1,MAXIT
C-----UPDATE TEMPERATURES
      CALL CALCT
C-----UPDATE MATERIAL PROPERTIES
      IF(INPRO) CALL PROPS
C-----UPDATE BOUNDARY CONDITIONS AND SOURCES IF NECESSARY
C
C-----CALCULATE NORMALISED RESIDUAL SOURCE
      SOURCE=RESORT/SNORM
C-----PRINT OUT ITERATION MONITORING INFORMATION
      WRITE(6,3700)NITER,SOURCE,T(IMON,JMON),TIME,DT,NSTEP
C-----OUTPUT TEMPERATURES AT INTERVALS SPECIFIED BY NITPRI
      IF(MOD(NITER,NITPRI).NE.0)GO TO 3300
      CALL PRINT(1,1,NI,NJ,IT,JT,X,Y,T,HEDT)
      IF (NSTEP.NE.MAXSTP.OR.SOURCE.GT.SORMAX) WRITE (6,3000) IMON,JMON
 3300 CONTINUE
```

```
C-----NORMAL EXIT FROM ITERATION LOOP WHEN SOURCE IS SUFFICIENTLY SMALL
      IF (SOURCE.LE.SORMAX) GO TO 3500
C-----EXIT FROM CALCULATIONS IF SOURCE IS LARGE AFTER 20 ITERATIONS
      IF (NITER.LE.20.OR.SOURCE.LT.100.) GO TO 3400
      WRITE (6,3800)
      GO TO 4100
 3400 CONTINUE
C--------------------END OF ITERATION LOOP------------------------------
C-----TERMINATE CALCULATIONS IF ITERATION DOES NOT CONVERGE
      WRITE (6,3900)
      GO TO 4100
C-----OUTPUT CONVERGED SOLUTION AT INTERVALS SPECIFIED BY NSTPRI
 3500 IF (MOD(NSTEP,NSTPRI).NE.0.OR.MOD(NITER,NITPRI).EQ.0) GO TO 3600
      CALL PRINT(1,1,NI,NJ,IT,JT,X,Y,T,HEDT)
      IF(NSTEP.NE.MAXSTP)WRITE(6,3000)IMON,JMON
 3600 CONTINUE
C--------------------END OF TIME LOOP-----------------------------------
C
CHAPTER  4  4  4  4  4  4  FINAL OPERATIONS AND OUTPUT  4  4  4  4  4  4
C
 4100 CONTINUE
      IF (MOD(NITER,NITPRI).NE.0.AND.MOD(MAXSTP,NSTPRI).NE.0)
     1   CALL PRINT (1,1,NI,NJ,IT,JT,X,Y,T,HEDT)
      STOP
C-----FORMAT STATEMENTS
 2900 FORMAT(/44X,9HTEACH-C  //15X,65HCONDUCTION IN RECTANGULAR BAR WITH
     1 PRESCRIBED SURFACE TEMPERATURE//
     2/16X,40H HEIGHT, H -----------------------------=,1PG10.3,2H M
     2/16X,40H WIDTH, W ------------------------------=,G10.3,2H M
     2/16X,40H SPECIFIC HEAT, CV ---------------------=,G10.3,7H J/KG K
     2/16X,40H THERMAL CONDUCTIVITY, TCON ------------=,G10.3,8H J/M S K
     2/16X,40H DENSITY, DENSIT -----------------------=,G10.3,8H KG/M**3
     2/16X,40H INITIAL TIME STEP, DT -----------------=,G10.3,2H S
     2/16X,40H SOURCE NORMALISATION FACTOR, CNORM ----=,G10.3
     2/16X,40H NUMBER OF NODES IN X DIRECTION, NI ----=,3X,I3
     2/16X,40H NUMBER OF NODES IN Y DIRECTION, NJ ----=,3X,I3)
 3000 FORMAT(//17X,5HNITER,3X,6HSOURCE,5X,2HT(,I2,1H,,I2,1H),4X,
     1       7HTIME(S),6X,5HDT(S),7X,5HNSTEP)
 3700 FORMAT(1H ,16X,I3,1P4E12.3,6X,I3)
 3800 FORMAT (//10X,46HSOURCE IS LARGE AFTER MORE THAN 20 ITERATIONS ,
     1       29H** CALCULATIONS TERMINATED **//)
 3900 FORMAT(//12X,49H** CAUTION ** CONVERGENCE CRITERION NOT SATISFIED
     1       /25X,39H WHEN PROGRAM TERMINATED AT NITER=MAXIT)
      END
C
C
```

```
      SUBROUTINE PROPS
C
C----------------CALCULATION OF MATERIAL PROPERTIES---------------------
C
CHAPTER  0  0  0  0  0  0  0  0  PRELIMINARIES  0  0  0  0  0  0  0  0
C
      COMMON
     1/PROP/TCON,CV,DENSIT,GAMH(22,22)
     1/GEOM/NI,NJ,NIM1,NJM1,INCYLX,INCYLY,DX,DY,RX(22),RU(22),
     1      X(22),Y(22),DXEP(22),DXPW(22),DYNP(22),DYPS(22),
     1      SNS(22),SEW(22),XU(22),YV(22),RY(22),RV(22),JS(22),JN(22)
C
CHAPTER 1  1  1  1  1  1  1  1  UPDATE PROPERTIES  1  1  1  1  1  1
C
      DO 1100 I=1,NI
      DO 1100 J=1,NJ
      GAMH(I,J)=TCON
 1100 CONTINUE
      RETURN
      END
C
C
```

```
      SUBROUTINE INIT
C
C--------------------------INITIALISATION OPERATIONS--------------------
C
CHAPTER  0  0  0  0  0  0  0  0  PRELIMINARIES  0  0  0  0  0  0  0  0
C
      COMMON
     1/CONVAR/IT,JT,INTIME,DT,RESORT,URFT,GREAT
     1/PROP/TCON,CV,DENSIT,GAMH(22,22)
     1/TEMP/T(22,22),TOLD(22,22)
     1/GEOM/NI,NJ,NIM1,NJM1,INCYLX,INCYLY,DX,DY,RX(22),RU(22),
     1      X(22),Y(22),DXEP(22),DXPW(22),DYNP(22),DYPS(22),
     1      SNS(22),SEW(22),XU(22),YV(22),RY(22),RV(22),JS(22),JN(22)
     1/COEF/IL,AP(22),AN(22),AS(22),AE(22),AW(22),SU(22),SP(22)
      LOGICAL INCYLX,INCYLY
C
CHAPTER  1  1  1  1  1  1  CALCULATE GRID QUANTITIES  1  1  1  1  1  1
C
      NIM1=NI-1
      NJM1=NJ-1
      IF (INCYLX.AND.INCYLY) GO TO 3100
C-----SET RX=X IF AXISYMMETRIC IN X DIRECTION
      DO 1100 I=1,NI
      RX(I)=1.0
 1100 IF (INCYLX) RX(I)=X(I)
C-----SET RY=Y IF AXISYMMETRIC IN Y DIRECTION
      DO 1200 J=1,NJ
      RY(J)=1.0
 1200 IF (INCYLY) RY(J)=Y(J)
C-----CALCULATE DISTANCES BETWEEN NODES
      DXPW(1)=0.0
      DXEP(NI)=0.0
      DO 1300 I=1,NIM1
      DXEP(I)=X(I+1)-X(I)
 1300 DXPW(I+1)=DXEP(I)
      DYPS(1)=0.0
      DYNP(NJ)=0.0
      DO 1400 J=1,NJM1
      DYNP(J)=Y(J+1)-Y(J)
 1400 DYPS(J+1)=DYNP(J)
C-----CALCULATE CELL WIDTHS AND HEIGHTS
      SEW(1)=0.0
      SEW(NI)=0.0
      DO 1500 I=2,NIM1
 1500 SEW(I)=0.5*(DXEP(I)+DXPW(I))
      SNS(1)=0.0
      SNS(NJ)=0.0
      DO 1600 J=2,NJM1
 1600 SNS(J)=0.5*(DYNP(J)+DYPS(J))
C-----FIND CELL BOUNDARIES
      XU(1)=0.0
      RU(1)=0.0
      DO 1700 I=2,NI
      RU(I)=0.5*(RX(I)+RX(I-1))
 1700 XU(I)=0.5*(X(I)+X(I-1))
      YV(1)=0.0
      RV(1)=0.0
      DO 1800 J=2,NJ
      RV(J)=0.5*(RY(J)+RY(J-1))
 1800 YV(J)=0.5*(Y(J)+Y(J-1))
C-----MODIFY BOUNDARY VALUES OF X AND Y
      X(1)=XU(2)
      IF (X(1).LT.(XU(NI)-XU(2))*1.E-3) X(1)=0.0
      X(NI)=XU(NI)
      Y(1)=YV(2)
      IF (Y(1).LT.(YV(NJ)-YV(2))*1.E-3) Y(1)=0.0
      Y(NJ)=YV(NJ)
C
CHAPTER  2  2  2  2  2  2  2  SET ARRAYS TO ZERO  2  2  2  2  2  2  2
C
      DO 2100 J=1,NJ
      AE(J)=0.0
      AW(J)=0.0
      AN(J)=0.0
      AS(J)=0.0
      SU(J)=0.0
      SP(J)=0.0
      DO 2100 I=1,NI
      GAMH(I,J)=0.0
      TOLD(I,J)=0.0
 2100 T(I,J)=0.0
      RETURN
```

```
C                                                                         INIT
CHAPTER  3  3  3  INPUT ERROR WARNINGS AND PROGRAM TERMINATION  3  3  3   INIT
C                                                                         INIT
C-----TERMINATE PROGRAM IF INCYLX AND INCYLY =.TRUE.                      INIT
 3100 WRITE (6,3900)                                                      INIT
      STOP                                                                INIT
 3900 FORMAT (//10X,37H INCYLX AND INCYLY BOTH SET TO .TRUE.              INIT
     1    //16X,26H*** PROGRAM TERMINATED ***)                            INIT
      END                                                                 INIT
C                                                                         INIT
C                                                                         INIT
```

```
      SUBROUTINE CALCT                                                    CALCT
C                                                                         CALCT
C----------------------SOLUTION OF ENERGY EQUATION---------------------  CALCT
C                                                                         CALCT
CHAPTER  0  0  0  0  0  0  0  PRELIMINARIES  0  0  0  0  0  0  0          CALCT
C                                                                         CALCT
      COMMON                                                              CALCT
     1/CONVAR/IT,JT,INTIME,DT,RESORT,URFT,GREAT                           CALCT
     1/PROP/TCON,CV,DENSIT,GAMH(22,22)                                    CALCT
     1/TEMP/T(22,22),TOLD(22,22)                                          CALCT
     1/GEOM/NI,NJ,NIM1,NJM1,INCYLX,INCYLY,DX,DY,RX(22),RU(22),            CALCT
     1      X(22),Y(22),DXEP(22),DXPW(22),DYNP(22),DYPS(22),              CALCT
     1      SNS(22),SEW(22),XU(22),YV(22),RY(22),RV(22),JS(22),JN(22)     CALCT
     1/COEF/IL,AP(22),AN(22),AS(22),AE(22),AW(22),SU(22),SP(22)           CALCT
      LOGICAL INTIME                                                      CALCT
C                                                                         CALCT
CHAPTER  1  1  1  1  1  1  ASSEMBLY OF COEFFICIENTS  1  1  1  1  1  1     CALCT
C                                                                         CALCT
      RESORT=0.0                                                          CALCT
C-----LOOP ALONG LINES                                                    CALCT
      DO 4200 I=2,NIM1                                                    CALCT
C-----FIND TRAVERSE LIMITS FOR THIS LINE                                  CALCT
      LJS=JS(I)                                                           CALCT
      LJN=JN(I)                                                           CALCT
C-----CALCULATE MAIN COEFFICIENTS FOR WHOLE LINE                          CALCT
      DO 1200 J=LJS,LJN                                                   CALCT
C-----COMPUTE AREAS AND VOLUME                                            CALCT
      AREAN=RV(J+1)*SEW(I)*RX(I)                                          CALCT
      AREAE=RY(J)*SNS(J)*RU(I+1)                                          CALCT
      VOL=RY(J)*SNS(J)*SEW(I)*RX(I)                                       CALCT
C-----CALCULATE DIFFUSION COEFFICIENTS                                    CALCT
      GAMN=0.5*(GAMH(I,J)+GAMH(I,J+1))                                    CALCT
      GAME=0.5*(GAMH(I,J)+GAMH(I+1,J))                                    CALCT
      DN=GAMN*AREAN/DYNP(J)                                               CALCT
      DE=GAME*AREAE /DXEP(I)                                              CALCT
C-----SOURCE TERMS                                                        CALCT
      SU(J)=0.0                                                           CALCT
      SP(J)=0.0                                                           CALCT
      IF(.NOT.INTIME)GO TO 1100                                           CALCT
      D=DENSIT*CV*VOL/DT                                                  CALCT
      SU(J)=D*TOLD(I,J)                                                   CALCT
      SP(J)=-D                                                            CALCT
 1100 CONTINUE                                                            CALCT
C-----ASSEMBLE MAIN COEFFICIENTS                                          CALCT
      AN(J)=DN                                                            CALCT
      AS(J)=AN(J-1)                                                       CALCT
      AW(J)=AE(J)                                                         CALCT
      AE(J)=DE                                                            CALCT
 1200 CONTINUE                                                            CALCT
      IL=I                                                                CALCT
C                                                                         CALCT
CHAPTER  2  2  2  2  2  2  PROBLEM MODIFICATIONS  2  2  2  2  2  2  2     CALCT
C                                                                         CALCT
      CALL PROMOD                                                         CALCT
C                                                                         CALCT
CHAPTER 3  FINAL COEFFICIENT ASSEMBLY | RESIDUAL SOURCE CALCULATION       CALCT
C                                                                         CALCT
      DO 3100 J=LJS,LJN                                                   CALCT
      AP(J)=AN(J)+AS(J)+AE(J)+AW(J)-SP(J)                                 CALCT
      RESOR=AN(J)*T(I,J+1)+AS(J)*T(I,J-1)+AE(J)*T(I+1,J)                  CALCT
     1     +AW(J)*T(I-1,J)-AP(J)*T(I,J)+SU(J)                             CALCT
      VOL=RY(J)*SEW(I)*SNS(J)*RX(I)                                       CALCT
C-----MODIFY RESOR WHEN BOUNDARY CONDITIONS ARE                           CALCT
C-----APPLIED USING SP=-GREAT.                                            CALCT
```

```
      IF (-SP(J).GT.0.5*GREAT) RESOR=RESOR/GREAT                       CALCT
C-----ADD RESIDUAL SOURCE FOR THIS LINE                                CALCT
      RESORT=RESORT+ABS(RESOR)                                         CALCT
C-----------UNDER-RELAXATION                                           CALCT
      AP(J)=AP(J)/URFT                                                 CALCT
      SU(J)=SU(J)+(1.-URFT)*AP(J)*T(I,J)                               CALCT
 3100 CONTINUE                                                         CALCT
C                                                                      CALCT
CHAPTER  4  4  4  4  4  SOLUTION OF DIFFERENCE EQUATIONS  4  4  4  4  4 CALCT
C                                                                      CALCT
C-----PERFORM LINE ITERATION                                           CALCT
 4100 CALL LISOLV(LJS,LJN+1,IT,JT,T)                                   CALCT
 4200 CONTINUE                                                         CALCT
C-----PERFORM BLOCK ADJUSTMENTS                                        CALCT
      IF (NI.GT.3) CALL BLKSLV (2,NI,IT,JT,T)                          CALCT
      RETURN                                                           CALCT
      END                                                              CALCT
C                                                                      CALCT
C                                                                      CALCT
```

```
      SUBROUTINE SOLVE (JSTART,JEND,IT,JT,PHI)                         SOLVE
C                                                                      SOLVE
C------------- SOLUTION OF EQUATIONS BY ITERATION ---------------       SOLVE
C                                                                      SOLVE
CHAPTER  0  0  0  0  0  0  0  0  PRELIMINARIES  0  0  0  0  0  0  0  0  SOLVE
C                                                                      SOLVE
      DIMENSION AB(22),BB(22),CB(22),DB(22)                            SOLVE
      DIMENSION PHI(IT,JT).A(22),B(22),C(22),D(22)                     SOLVE
      COMMON                                                           SOLVE
     1/GEOM/NI,NJ,NIM1,NJM1,INCYLX,INCYLY,DX,DY,RX(22),RU(22),         SOLVE
     1      X(22),Y(22),DXEP(22),DXPW(22),DYNP(22),DYPS(22),           SOLVE
     1      SNS(22),SEW(22),XU(22);YV(22),RY(22),RV(22),JS(22),JN(22)  SOLVE
     1/COEF/IL,AP(22),AN(22),AS(22),AE(22),AW(22),SU(22),SP(22)        SOLVE
C                                                                      SOLVE
CHAPTER  1  1  1  1  1  LINE ITERATION PROCEDURE  1  1  1  1  1  1  1  SOLVE
C                                                                      SOLVE
      ENTRY LISOLV                                                     SOLVE
      JENDM1=JEND-1                                                    SOLVE
      JSTM1=JSTART-1                                                   SOLVE
      A(JSTM1)=0.0                                                     SOLVE
      I=IL                                                             SOLVE
      C(JSTM1)=PHI(I,JSTM1)                                            SOLVE
C-----COMMENCE TRAVERSE                                                SOLVE
      DO 1100 J=JSTART,JENDM1                                          SOLVE
C-----ASSEMBLE TDMA COEFFICIENTS                                       SOLVE
      A(J)=AN(J)                                                       SOLVE
      B(J)=AS(J)                                                       SOLVE
      C(J)=AE(J)*PHI(I+1,J)+AW(J)*PHI(I-1,J)+SU(J)                     SOLVE
      D(J)=AP(J)                                                       SOLVE
C-----CALCULATE COEFFICIENTS OF RECURRENCE FORMULA                     SOLVE
      TERM=1./(D(J)-B(J)*A(J-1))                                       SOLVE
      A(J)=A(J)*TERM                                                   SOLVE
 1100 C(J)=(C(J)+B(J)*C(J-1))*TERM                                     SOLVE
C-----OBTAIN NEW PHI'S BY BACK SUBSTITUTION                            SOLVE
      DO 1200 JJ=JSTART,JENDM1                                         SOLVE
      J=JEND+JSTM1-JJ                                                  SOLVE
 1200 PHI(I,J)=A(J)*PHI(I,J+1)+C(J)                                    SOLVE
      IF (NI.LE.3) GO TO 1600                                          SOLVE
C-----------SEQUENCE FOR CALCULATING BLOCK ADJUSTMENT COEFFICIENTS----- SOLVE
C-----SET BLOCK COEFFICIENTS IF I IS 2                                 SOLVE
      IF (I.NE.2) GO TO 1400                                           SOLVE
      DO 1300 K=1,NI                                                   SOLVE
      AB(K)=0.0                                                        SOLVE
      BB(K)=0.0                                                        SOLVE
      CB(K)=0.0                                                        SOLVE
      DB(K)=0.0                                                        SOLVE
 1300 CONTINUE                                                         SOLVE
C-----ASSEMBLE BLOCK COEFFICIENTS                                      SOLVE
 1400 DO 1500 J=JSTART,JENDM1                                          SOLVE
      AB(I)=AB(I)+AE(J)                                                SOLVE
      BB(I)=BB(I)+AW(J)                                                SOLVE
      CB(I)=CB(I)+AW(J)*PHI(I-1,J)-(AE(J)+AW(J)-SP(J))*PHI(I,J)+SU(J)  SOLVE
      CB(I-1)=CB(I-1)+AW(J)*PHI(I,J)                                   SOLVE
      DB(I)=DB(I)+AE(J)+AW(J)-SP(J)                                    SOLVE
 1500 CONTINUE                                                         SOLVE
 1600 RETURN                                                           SOLVE
C                                                                      SOLVE
```

```
CHAPTER  2  2  2  2  2  BLOCK ADJUSTMENT PROCEDURE  2  2  2  2  2  2
C
      ENTRY BLKSLV
      JENDM1=JEND-1
      JSTM1=JSTART-1
      AB(JSTM1)=0.0
      CB(JSTM1)=0.0
C-----CALCULATE COEFFICIENTS OF RECURRENCE FORMULA
      DO 2100 J=JSTART,JENDM1
      TERM=1./(DB(J)-BB(J)*AB(J-1))
      AB(J)=AB(J)*TERM
 2100 CB(J)=(CB(J)+BB(J)*CB(J-1))*TERM
C-----OBTAIN BLOCK ADJUSTMENTS AND STORE IN BB
      BB(JEND)=0.0
      DO 2200 JJ=JSTART,JENDM1
      J=JEND+JSTM1-JJ
 2200 BB(J)=AB(J)*BB(J+1)+CB(J)
C-----ADD ADJUSTMENTS TO EACH LINE
      DO 2300 I=JSTART,JENDM1
      ADJ=BB(I)
      LJS=JS(I)
      LJN=JN(I)
      DO 2300 J=LJS,LJN
      PHI(I,J)=PHI(I,J)+ADJ
 2300 CONTINUE
      RETURN
      END
C
C
```

```
      SUBROUTINE PROMOD
C
C------------------------- PROBLEM MODIFICATIONS ----------------------
C
CHAPTER  0  0  0  0  0  0  0  PRELIMINARIES  0  0  0  0  0  0  0  0  0
C
      COMMON
     1/CONVAR/IT,JT,INTIME,DT,RESORT,URFT,BREAT
     1/PROP/TCON,CV,DENSIT,GAMH(22,22)
     1/TEMP/T(22,22),TOLD(22,22)
     1/GEOM/NI,NJ,NIM1,NJM1,INCYLX,INCYLY,DX,DY,RX(22),RU(22),
     1      X(22),Y(22),DXEP(22),DXPW(22),DYNP(22),DYPS(22),
     1      SNS(22),SEW(22),XU(22),YV(22),RY(22),RV(22),JS(22),JN(22)
     1/COEF/IL,AP(22),AN(22),AS(22),AE(22),AW(22),SU(22),SP(22)
C
CHAPTER  1  1  1  1  1  1  1  THERMAL ENERGY  1  1  1  1  1  1  1  1  1
C
C--------------------------TOP AND BOTTOM BOUNDARIES------------------
      RDYN=RV(NJ)/(YV(NJ)-Y(NJM1))
      RDYS=RV(2)/(Y(2)-YV(2))
C-----SET NORTH COEFFICIENT TO ZERO
      AN(NJM1)=0.0
C-----CALCULATE DIFFUSION COEFFICIENT AT NORTH BOUNDARY
      DN=GAMH(IL,NJM1)*SEW(IL)*RDYN
C-----INSERT CORRECT BOUNDARY CONDITION VIA SU AND SP
      SU(NJM1)=SU(NJM1)+DN*T(IL,NJ)
      SP(NJM1)=SP(NJM1)-DN
      AS(2)=0.0
      DS=GAMH(IL,2)*SEW(IL)*RDYS
      SU(2)=SU(2)+DS*T(IL,1)
      SP(2)=SP(2)-DS
C-----------------------------SIDE BOUNDARIES-------------------------
      IF (IL.NE.NIM1) GO TO 1200
      DXE=XU(NI)-X(NIM1)
      DO 1100 J=2,NJM1
      AE(J)=0.0
      DE=GAMH(IL,J)*SNS(J)*RY(J)/DXE
      SU(J)=SU(J)+DE*T(NI,J)
      SP(J)=SP(J)-DE
 1100 CONTINUE
 1200 IF (IL.NE.2) GO TO 1400
      DXW=X(2)-XU(2)
      DO 1300 J=2,NJM1
      AW(J)=0.0
      DW=GAMH(IL,J)*SNS(J)*RY(J)/DXW
      SU(J)=SU(J)+DW*T(1,J)
      SP(J)=SP(J)-DW
 1300 CONTINUE
 1400 RETURN
      END
C
C
```

```
      SUBROUTINE PRINT(ISTART,JSTART,IEND,JEND,IT,JT,X,Y,PHI,HEAD)
C
C------------------ OUTPUT OF DEPENDENT-VARIABLE FIELDS ----------------
C
CHAPTER  0  0  0  0  0  0  0  0  PRELIMINARIES  0  0  0  0  0  0  0  0
C
      DIMENSION PHI(IT,JT),X(IT),Y(JT),HEAD(6),STORE(12)
      DIMENSION F(8),F4(13),FD(7)
      DATA F(1),F(2),F(3),F(4),F(5),F(6),F(7),F(8)
     1     /4H(/1H,4H ,A6,4H,I3,,4H7X, ,4H  11,4H(I3,,4H7X),,4HA6) /
      DATA FD(1),FD(2),FD(3),FD(4),FD(5),FD(6),FD(7)
     1     /4H(1H ,4H,I3,,4H  12,4H(1PE,4H10.2,4H),G1,4H1.3)/
      DATA F4(1),F4(2),F4(3),F4(4),F4(5),F4(6),F4(7)
     1     ,F4(8),F4(9),F4(10),F4(11),F4(12),F4(13)
     1     /4HA6, ,4H   1,4H   2,4H   3,4H   4,4H   5,
     1       4H   6,4H   7,4H   8,4H   9,4H  10,4H  11,4H  12/
      DATA HI,HY/6H   I =, 6HY =   /
C
CHAPTER  1  1  1  1  1  1  INITIALIZATION AND HEADINGS  1  1  1  1  1
C
      ISKIP=1
      JSKIP=1
      LINLIM=12
      LINSTA=ISTART
C-----PRINT ARRAY HEADING
      WRITE(6,1800)HEAD
 1100 CONTINUE
      LINEND=LINSTA+(LINLIM-1)*ISKIP
      LINEND=MINO(IEND,LINEND)
      IF4=(LINEND-LINSTA)/ISKIP+1
      F(5)=F4(IF4)
      FD(3)=F4(IF4+1)
C-----PRINT LINE HEADINGS
      WRITE(6,F) HI, (I,I=LINSTA,LINEND,ISKIP), HY
      WRITE(6,1900)
C
CHAPTER  2  2  2  2  2  2  OUTPUT OF PHI ARRAY  2  2  2  2  2  2  2  2
C
      DO 2200 JJ=JSTART,JEND,JSKIP
      J=JSTART+JEND-JJ
      IS=0
      DO 2100 I=LINSTA,LINEND,ISKIP
      A=PHI(I,J)
      IF(ABS(A).LT.1.E-20) A=0.0
      IS=IS+1
 2100 STORE(IS)=A
 2200 WRITE(6,FD)  J,(STORE(I),I=1,IS),Y(J)
      WRITE(6,2900) (X(I),I=LINSTA,LINEND,ISKIP)
      LINSTA=LINEND+ISKIP
      IF(LINEND.LT.IEND)GO TO 1100
      RETURN
 1800 FORMAT(//1X,20(2H*-),7X,6A6,7X,20(2H-*))
 1900 FORMAT(3H  J)
 2900 FORMAT(/4H X =,1PG10.3,11G10.3)
      END
```

APPENDIX D – Output from TEACH–C

```
                                 TEACH-C

   CONDUCTION IN RECTANGULAR BAR WITH PRESCRIBED SURFACE TEMPERATURE

      HEIGHT, H -----------------------------= 1.00      M
      WIDTH, W ------------------------------= 1.00      M
      SPECIFIC HEAT, CV --------------------= 460.      J/KG K
      THERMAL CONDUCTIVITY, TCON -----------= 52.0      J/M S K
      DENSITY, DENSIT ----------------------= 7.850E+03 KG/M**3
      INITIAL TIME STEP, DT ----------------= 50.0      S
      SOURCE NORMALISATION FACTOR, SNORM ---= 5.200E+03
      NUMBER OF NODES IN X DIRECTION, NI ---=    12
      NUMBER OF NODES IN Y DIRECTION, NJ ---=    12
```

- TEMPERATURE (C) -

J	I = 1	2	3	4	5	6	7	8	9	10	11	12	Y =
12	0	1.00E+02	1.00E+02	1.00E+02	1.00E+02	1.00E+02	1.00E+02	1.00E+02	1.00E+02	1.00E+02	1.00E+02	0	1.00
11	0	0	0	0	0	0	0	0	0	0	0	0	.950
10	0	0	0	0	0	0	0	0	0	0	0	0	.850
9	0	0	0	0	0	0	0	0	0	0	0	0	.750
8	0	0	0	0	0	0	0	0	0	0	0	0	.650
7	0	0	0	0	0	0	0	0	0	0	0	0	.550
6	0	0	0	0	0	0	0	0	0	0	0	0	.450
5	0	0	0	0	0	0	0	0	0	0	0	0	.350
4	0	0	0	0	0	0	0	0	0	0	0	0	.250
3	0	0	0	0	0	0	0	0	0	0	0	0	.150
2	0	0	0	0	0	0	0	0	0	0	0	0	5.000E
1	0	0	0	0	0	0	0	0	0	0	0	0	
X =	0	5.000E-02	.150	.250	.350	.450	.550	.650	.750	.850	.950	1.00	

NITER	SOURCE	T(6, 6)	TIME(S)	DT(S)	NSTEP
1	2.106E+01	8.348E-02	5.000E+01	5.000E+01	1
2	1.873E+00	5.171E-03	5.000E+01	5.000E+01	1
3	9.600E-02	3.328E-04	5.000E+01	5.000E+01	1
4	5.112E-03	3.194E-05	5.000E+01	5.000E+01	1
5	2.681E-04	1.313E-05	5.000E+01	5.000E+01	1
6	1.377E-05	1.200E-05	5.000E+01	5.000E+01	1

- TEMPERATURE (C) -

J	I = 1	2	3	4	5	6	7	8	9	10	11	12	Y =
12	0	1.00E+02	1.00E+02	1.00E+02	1.00E+02	1.00E+02	1.00E+02	1.00E+02	1.00E+02	1.00E+02	1.00E+02	0	1.00
11	0	1.07E+01	1.18E+01	1.19E+01	1.19E+01	1.19E+01	1.19E+01	1.19E+01	1.19E+01	1.18E+01	1.07E+01	0	.950
10	0	6.06E-01	7.39E-01	7.50E-01	7.51E-01	7.51E-01	7.51E-01	7.51E-01	7.50E-01	7.39E-01	6.06E-01	0	.850
9	0	3.47E-02	4.61E-02	4.73E-02	4.75E-02	4.75E-02	4.75E-02	4.75E-02	4.73E-02	4.61E-02	3.47E-02	0	.750
8	0	[illegible]	[illegible]	2.90E-03	3.00E-03	3.00E-03	3.00E-03	3.00E-03	2.90E-03	2.86E-03	1.99E-03	0	.650
7	0	1.15E-04	1.70E-04	1.88E-04	1.89E-04	1.90E-04	[illegible]	[illegible]	1.88E-04	1.70E-04	1.15E-04	0	.550
6	0	6.77E-06	1.11E-05	1.19E-05	1.20E-05	1.20E-05	1.20E-05	1.20E-05	1.18E-05	1.10E-05	6.70E-06	0	.450
5	0	4.62E-07	7.60E-07	8.24E-07	8.29E-07	7.79E-07	7.61E-07	7.56E-07	7.45E-07	6.81E-07	3.92E-07	0	.350
4	0	9.23E-08	1.21E-07	1.25E-07	1.20E-07	6.99E-08	5.20E-08	4.83E-08	4.69E-08	4.21E-08	2.31E-08	0	.250
3	0	6.55E-08	7.51E-08	7.56E-08	7.02E-08	2.31E-08	6.72E-09	3.51E-09	2.99E-09	2.61E-09	1.36E-09	0	.150
2	0	3.63E-08	4.07E-08	4.08E-08	3.76E-08	1.05E-08	1.91E-09	3.93E-10	1.94E-10	1.53E-10	7.62E-11	0	5.000E
1	0	0	0	0	0	0	0	0	0	0	0	0	
X =	0	5.000E-02	.150	.250	.350	.450	.550	.650	.750	.850	.950	1.00	

NITER	SOURCE	T(6, 6)	TIME(S)	DT(S)	NSTEP
1	1.815E+01	7.416E-02	1.000E+02	5.000E+01	2
2	1.548E+00	4.687E-03	1.000E+02	5.000E+01	2
3	8.355E-02	3.632E-04	1.000E+02	5.000E+01	2
4	4.454E-03	9.259E-05	1.000E+02	5.000E+01	2
5	2.338E-04	7.564E-05	1.000E+02	5.000E+01	2
6	1.200E-05	7.469E-05	1.000E+02	5.000E+01	2

- TEMPERATURE (C) -

J	I = 1	2	3	4	5	6	7	8	9	10	11	12	Y =
12	0	1.00E+02	1.00E+02	1.00E+02	1.00E+02	1.00E+02	1.00E+02	1.00E+02	1.00E+02	1.00E+02	1.00E+02	0	1.00
11	0	1.87E+01	2.15E+01	2.17E+01	2.17E+01	2.17E+01	2.17E+01	2.17E+01	2.17E+01	2.15E+01	1.87E+01	0	.950
10	0	1.55E+00	1.99E+00	2.03E+00	2.04E+00	2.04E+00	2.04E+00	2.04E+00	2.03E+00	1.99E+00	1.55E+00	0	.850
9	0	1.16E-01	1.64E-01	1.70E-01	1.70E-01	1.70E-01	1.70E-01	1.70E-01	1.70E-01	1.64E-01	1.16E-01	0	.750
8	0	8.33E-03	1.26E-02	1.33E-02	1.34E-02	1.34E-02	1.34E-02	1.34E-02	1.33E-02	1.26E-02	8.33E-03	0	.650
7	0	5.77E-04	9.36E-04	1.00E-03	1.01E-03	1.01E-03	1.01E-03	1.01E-03	1.00E-03	9.36E-04	5.77E-04	0	.550
6	0	3.93E-05	6.76E-05	7.37E-05	7.46E-05	7.47E-05	7.47E-05	7.45E-05	7.35E-05	6.74E-05	3.92E-05	0	.450
5	0	2.75E-06	4.91E-06	5.43E-06	5.50E-06	5.43E-06	5.39E-06	5.37E-06	5.27E-06	4.76E-06	2.62E-06	0	.350
4	0	2.99E-07	4.79E-07	5.22E-07	5.12E-07	4.25E-07	3.91E-07	3.82E-07	3.73E-07	3.31E-07	1.74E-07	0	.250
3	0	1.26E-07	1.59E-07	1.62E-07	1.47E-07	6.50E-08	3.44E-08	2.78E-08	2.61E-08	2.27E-08	1.15E-08	0	.150
2	0	6.43E-08	7.63E-08	7.65E-08	6.73E-08	2.12E-08	5.32E-09	2.22E-09	1.72E-09	1.44E-09	7.02E-10	0	5.000E
1	0	0	0	0	0	0	0	0	0	0	0	0	
X =	0	5.000E-02	.150	.250	.350	.450	.550	.650	.750	.850	.950	1.00	

NITER	SOURCE	T(6, 6)	TIME(S)	DT(S)	NSTEP
1	1.579E+01	6.654E-02	1.500E+02	5.000E+01	3
2	1.348E+00	4.420E-03	1.500E+02	5.000E+01	3
3	7.299E-02	5.273E-04	1.500E+02	5.000E+01	3
4	3.897E-03	2.824E-04	1.500E+02	5.000E+01	3
5	2.047E-04	2.671E-04	1.500E+02	5.000E+01	3
6	1.051E-05	2.663E-04	1.500E+02	5.000E+01	3

- TEMPERATURE (C) -

J	I = 1	2	3	4	5	6	7	8	9	10	11	12	Y =
12	0	1.00E+02	1.00E+02	1.00E+02	1.00E+02	1.00E+02	1.00E+02	1.00E+02	1.00E+02	1.00E+02	1.00E+02	0	1.00
11	0	2.47E+01	2.95E+01	2.99E+01	2.99E+01	2.99E+01	2.99E+01	2.99E+01	2.99E+01	2.95E+01	2.47E+01	0	.950
10	0	2.65E+00	3.56E+00	3.68E+00	3.69E+00	3.69E+00	3.69E+00	3.69E+00	3.68E+00	3.56E+00	2.65E+00	0	.850
9	0	2.46E-01	3.63E-01	3.81E-01	3.83E-01	3.83E-01	3.83E-01	3.83E-01	3.81E-01	3.63E-01	2.46E-01	0	.750
8	0	2.10E-02	3.34E-02	3.57E-02	3.60E-02	3.60E-02	3.60E-02	3.60E-02	3.57E-02	3.34E-02	2.10E-02	0	.650
7	0	1.70E-03	2.88E-03	3.13E-03	3.17E-03	3.17E-03	3.17E-03	3.17E-03	3.13E-03	2.88E-03	1.70E-03	0	.550
6	0	[illegible]	[illegible]	2.61E-04	2.66E-04	2.66E-04	2.66E-04	2.66E-04	2.61E-04	2.36E-04	1.32E-04	0	.450
5	0	1.01E-05	1.89E-05	2.12E-05	2.17E-05	[illegible]	[illegible]	[illegible]	2.10E-05	1.87E-05	9.93E-06	0	.350
4	0	9.02E-07	1.65E-06	1.86E-06	1.87E-06	1.76E-06	1.71E-06	1.69E-06	1.65E-06	1.44E-06	7.33E-07	0	[illegible]
3	0	2.07E-07	3.01E-07	3.18E-07	2.93E-07	1.85E-07	1.43E-07	1.32E-07	1.26E-07	1.09E-07	5.32E-08	0	.150
2	0	8.80E-08	1.11E-07	1.12E-07	9.61E-08	3.66E-08	1.47E-08	9.94E-09	8.88E-09	7.47E-09	3.54E-09	0	5.000E
1	0	0	0	0	0	0	0	0	0	0	0	0	
X =	0	5.000E-02	.150	.250	.350	.450	.550	.650	.750	.850	.950	1.00	

NITER	SOURCE	T(6, 6)	TIME(S)	DT(S)	NSTEP
1	1.387E+01	6.044E-02	2.000E+02	5.000E+01	4

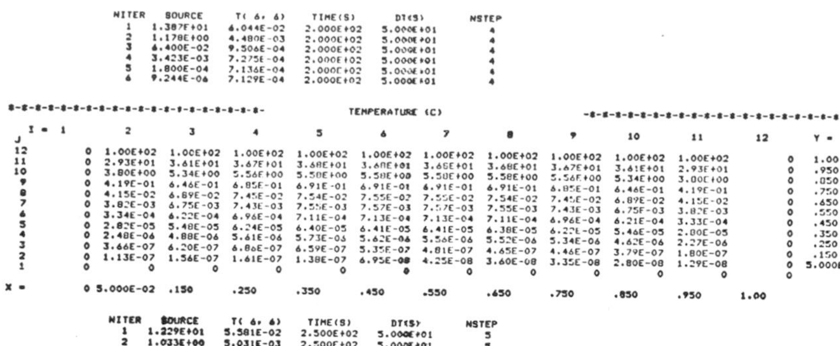

NITER	SOURCE	T(6, 6)	TIME(S)	DT(S)	NSTEP
1	1.387E+01	6.044E-02	2.000E+02	5.000E+01	4
2	1.178E+00	4.480E-03	2.000E+02	5.000E+01	4
3	6.400E-02	9.506E-04	2.000E+02	5.000E+01	4
4	3.423E-03	7.275E-04	2.000E+02	5.000E+01	4
5	1.800E-04	7.136E-04	2.000E+02	5.000E+01	4
6	9.244E-06	7.129E-04	2.000E+02	5.000E+01	4

TEMPERATURE (C)

J	I = 1	2	3	4	5	6	7	8	9	10	11	12	Y =
12	0	1.00E+02	1.00E+02	1.00E+02	1.00E+02	1.00E+02	1.00E+02	1.00E+02	1.00E+02	1.00E+02	1.00E+02	0	1.00
11	0	2.93E+01	3.61E+01	3.67E+01	3.68E+01	3.68E+01	3.65E+01	3.68E+01	3.67E+01	3.61E+01	2.93E+01	0	.950
10	0	3.80E+00	5.34E+00	5.56E+00	5.58E+00	5.58E+00	5.50E+00	5.58E+00	5.56E+00	5.34E+00	3.00E+00	0	.850
9	0	4.19E-01	6.46E-01	6.85E-01	6.91E-01	6.91E-01	6.91E-01	6.91E-01	6.85E-01	6.46E-01	4.19E-01	0	.750
8	0	4.15E-02	6.89E-02	7.45E-02	7.54E-02	7.55E-02	7.55E-02	7.54E-02	7.45E-02	6.89E-02	4.15E-02	0	.650
7	0	3.82E-03	6.75E-03	7.43E-03	7.55E-03	7.57E-03	7.57E-03	7.55E-03	7.43E-03	6.75F-03	3.82E-03	0	.550
6	0	3.34E-04	6.22E-04	6.96E-04	7.11E-04	7.13E-04	7.13E-04	7.11E-04	6.96E-04	6.21E-04	3.33E-04	0	.450
5	0	2.82E-05	5.48E-05	6.24E-05	6.40E-05	6.41E-05	6.41E-05	6.38E-05	6.22E-05	5.46E-05	2.80E-05	0	.350
4	0	2.48E-06	4.88E-06	5.61E-06	5.73E-06	5.62E-06	5.56E-06	5.52E-06	5.34E-06	4.62E-06	2.27E-06	0	.250
3	0	3.66E-07	6.20E-07	6.86E-07	6.59E-07	5.35E-07	4.81E-07	4.65E-07	4.46E-07	3.79E-07	1.80E-07	0	.150
2	0	1.13E-07	1.56E-07	1.61E-07	1.38E-07	6.95E-08	4.25E-08	3.60E-08	3.35E-08	2.80E-08	1.29E-08	0	5.000E
1	0	0	0	0	0	0	0	0	0	0	0	0	
X =	0	5.000E-02	.150	.250	.350	.450	.550	.650	.750	.850	.950	1.00	

NITER	SOURCE	T(6, 6)	TIME(S)	DT(S)	NSTEP
1	1.229E+01	5.581E-02	2.500E+02	5.000E+01	5
2	1.033E+00	5.031E-03	2.500E+02	5.000E+01	5

and so on.....

TEMPERATURE (C)

J	I = 1	2	3	4	5	6	7	8	9	10	11	12	Y =
12	0	1.00E+02	1.00E+02	1.00E+02	1.00E+02	1.00E+02	1.00E+02	1.00E+02	1.00E+02	1.00E+02	1.00E+02	0	1.00
11	0	4.62E+01	6.70E+01	7.16E+01	7.26E+01	7.27E+01	7.27E+01	7.26E+01	7.16E+01	6.70E+01	4.62E+01	0	.950
10	0	1.43E+01	2.67E+01	3.06E+01	3.17E+01	3.19E+01	3.19E+01	3.17E+01	3.06E+01	2.67E+01	1.43E+01	0	.850
9	0	4.14E+00	8.80E+00	1.06E+01	1.11E+01	1.12E+01	1.12E+01	1.11E+01	1.06E+01	8.80E+00	4.14E+00	0	.750
8	0	1.09E+00	2.49E+00	3.08E+00	3.29E+00	3.33E+00	3.33E+00	3.28E+00	3.08E+00	2.49E+00	1.09E+00	0	.650
7	0	2.58E-01	6.16E-01	7.81E-01	8.40E-01	8.56E-01	8.56E-01	8.40E-01	7.81E-01	6.16E-01	2.58E-01	0	.550
6	0	5.56E-02	1.37E-01	1.76E-01	1.91E-01	1.96E-01	1.96E-01	1.91E-01	1.76E-01	1.37E-01	5.56E-02	0	.450
5	0	1.10E-02	2.76E-02	3.62E-02	3.95E-02	4.06E-02	4.06E-02	3.95E-02	3.62E-02	2.76E-02	1.10E-02	0	.350
4	0	2.02E-03	5.15E-03	6.83E-03	7.52E-03	7.74E-03	7.73E-03	7.51E-03	6.82E-03	5.14E-03	2.02E-03	0	.250
3	0	3.48E-04	8.95E-04	1.20E-03	1.33E-03	1.37E-03	1.37E-03	1.32E-03	1.19E-03	8.91E-04	3.45E-04	0	.150
2	0	4.90E-05	1.26E-04	1.69E-04	1.88E-04	1.93E-04	1.93E-04	1.86E-04	1.67E-04	1.24E-04	4.75E-05	0	5.000E
1	0	0	0	0	0	0	0	0	0	0	0	0	
X =	0	5.000E-02	.150	.250	.350	.450	.550	.650	.750	.850	.950	1.00	

NITER	SOURCE	T(6, 6)	TIME(S)	DT(S)	NSTEP
1	4.413E+00	2.381E-01	9.500E+02	5.000E+01	19
2	3.023E-01	2.381E-01	9.500E+02	5.000E+01	19
3	1.790E-02	2.368E-01	9.500E+02	5.000E+01	19
4	1.020E-03	2.367E-01	9.500E+02	5.000E+01	19
5	5.615E-05	2.367E-01	9.500E+02	5.000E+01	19

TEMPERATURE (C)

J	I = 1	2	3	4	5	6	7	8	9	10	11	12	Y =
12	0	1.00E+02	1.00E+02	1.00E+02	1.00E+02	1.00E+02	1.00E+02	1.00E+02	1.00E+02	1.00E+02	1.00E+02	0	1.00
11	0	4.65E+01	6.76E+01	7.24E+01	7.34E+01	7.36E+01	7.36E+01	7.34E+01	7.24E+01	6.76E+01	4.65E+01	0	.950
10	0	1.47E+01	2.76E+01	3.18E+01	3.30E+01	3.33E+01	3.33E+01	3.30E+01	3.18E+01	2.76E+01	1.47E+01	0	.850
9	0	4.38E+00	9.42E+00	1.14E+01	1.20E+01	1.21E+01	1.21E+01	1.20E+01	1.14E+01	9.42E+00	4.38E+00	0	.750
8	0	1.19E+00	2.75E+00	3.44E+00	3.67E+00	3.73E+00	3.73E+00	3.67E+00	3.44E+00	2.75E+00	1.19E+00	0	.650
7	0	2.94E-01	7.08E-01	9.05E-01	9.76E-01	9.97E-01	9.97E-01	9.76E-01	9.05E-01	7.08E-01	2.94E-01	0	.550
6	0	6.58E-02	1.63E-01	2.12E-01	2.31E-01	2.37E-01	2.37E-01	2.31E-01	2.12E-01	1.63E-01	6.58E-02	0	.450
5	0	1.35E-02	3.42E-02	4.51E-02	4.95E-02	5.09E-02	5.09E-02	4.95E-02	4.51E-02	3.42E-02	1.35E-02	0	.350
4	0	2.50E-03	6.61E-03	8.83E-03	9.76E-03	1.02E-02	1.01E-02	9.76E-03	8.82E-03	6.61E-03	2.57E-03	0	.250
3	0	4.59E-04	1.19E-03	1.60E-03	1.70E-03	1.84E-03	1.84E-03	1.78E-03	1.60E-03	1.19E-03	4.56E-04	0	.150
2	0	6.61E-05	1.72E-04	2.32E-04	2.59E-04	2.68E-04	2.67E-04	2.57E-04	2.30E-04	1.69E-04	6.46E-05	0	5.000E
1	0	0	0	0	0	0	0	0	0	0	0	0	
X =	0	5.000E-02	.150	.250	.350	.450	.550	.650	.750	.850	.950	1.00	

NITER	SOURCE	T(6, 6)	TIME(S)	DT(S)	NSTEP
1	4.227E+00	3.030E-01	1.000E+03	5.000E+01	20
2	2.867E-01	2.841E-01	1.000E+03	5.000E+01	20
3	1.701E-02	2.828E-01	1.000E+03	5.000E+01	20
4	9.705E-04	2.827E-01	1.000E+03	5.000E+01	20
5	5.346E-05	2.827E-01	1.000E+03	5.000E+01	20

TEMPERATURE (C)

J	I = 1	2	3	4	5	6	7	8	9	10	11	12	Y =
12	0	1.00E+02	1.00E+02	1.00E+02	1.00E+02	1.00E+02	1.00E+02	1.00E+02	1.00E+02	1.00E+02	1.00E+02	0	1.00
11	0	4.67E+01	6.81E+01	7.31E+01	7.42E+01	7.44E+01	7.44E+01	7.42E+01	7.31E+01	6.81E+01	4.67E+01	0	.950
10	0	1.50E+01	2.85E+01	3.30E+01	3.43E+01	3.46E+01	3.46E+01	3.43E+01	3.30E+01	2.85E+01	1.50E+01	0	.850
9	0	4.62E+00	1.00E+01	1.22E+01	1.29E+01	1.30E+01	1.30E+01	1.29E+01	1.22E+01	1.00E+01	4.62E+00	0	.750
8	0	1.30E+00	3.03E+00	3.81E+00	4.08E+00	4.15E+00	4.15E+00	4.08E+00	3.81E+00	3.03E+00	1.30E+00	0	.650
7	0	3.32E-01	8.06E-01	1.04E+00	1.12E+00	1.15E+00	1.15E+00	1.12E+00	1.04E+00	8.06E-01	3.32E-01	0	.550
6	0	7.70E-02	1.92E-01	2.52E-01	2.75E-01	2.83E-01	2.83E-01	2.75E-01	2.52E-01	1.92E-01	7.70E-02	0	.450
5	0	1.64E-02	4.18E-02	5.55E-02	6.12E-02	6.30E-02	6.30E-02	6.12E-02	5.55E-02	4.18E-02	1.64E-02	0	.350
4	0	3.24E-03	8.37E-03	1.12E-02	1.25E-02	1.29E-02	1.29E-02	1.25E-02	1.12E-02	8.36E-03	3.24E-03	0	.250
3	0	5.96E-04	1.56E-03	2.11E-03	2.36E-03	2.44E-03	2.44E-03	2.35E-03	2.11E-03	1.55E-03	5.94E-04	0	.150
2	0	8.80E-05	2.30E-04	3.14E-04	3.51E-04	3.64E-04	3.63E-04	3.50E-04	3.11E-04	2.28E-04	8.64E-05	0	5.000E
1	0	0	0	0	0	0	0	0	0	0	0	0	
X =	0	5.000E-02	.150	.250	.350	.450	.550	.650	.750	.850	.950	1.00	

STOP

APPENDIX E

The Use of UPDATE

UPDATE, it should be stressed is a CDC copyright product. In developing TEACH–C and the PROBLEMs we found it very useful, and we have written this book using the terminology of UPDATE. Similar systems may be available on other machines.

1. The Identifiers

Each line of the code is given an identifier: this appears in the 'free' columns beyond 72 on the FORTRAN cards or lines. Thus the line NI=12 has identification CONTRO.57 (page 57).

2. Making changes

The blocks of changes (see, for example, page 103 below) must be headed by a line which states how the changes are themselves to be identified when they have been made. In this case, the header is

*IDENT,PROBLEM2

so that any new cards (lines) inserted by the PROBLEM2 UPDATE will be identified as PROBLEM2.1, PROBLEM2.2 and so on, in the order in which they were performed by the UPDATE.

Most changes are of the form 'Delete this card (line) and replace it with the following line(s)'. In the case of the PROBLEM2 UPDATE, we have:

*D,CONTRO.184,185

which means 'delete cards (lines) with identifiers CONTRO.184 to CONTRO.185 (inclusive!)'
followed by a line of FORTRAN which is inserted in the place of the deleted cards (lines) and which receives the identifier PROBLEM2.1 because it is the first card (line) inserted as a result of the PROBLEM2 UPDATE.

One other UPDATE instruction is occasionally used. It is the instruction: 'Simply insert the following cards (lines) after a given line': for example, on page 106 you will find in the P2L1 UPDATE the line: *I,CONTRO.133 followed by two lines of FORTRAN which must be inserted immediately *after* CONTRO.133, and which receive the identifiers P2L1.7 and P2L1.8 respectively, as they are the seventh and eighth lines inserted as a result of the P2L1 UPDATE. In any subsequent UPDATE (for a RUN) they would have to be referred to as P2L1.7 and P2L1.8.

Applications of TEACH–C

PROBLEM 1

Unsteady One-dimensional Conduction Processes

1. Introduction

Both in engineering and in the natural environment there are many important problems of time-dependent heat conduction in which substantial gradients in temperature occur predominantly in one coordinate direction. Some examples are:

- the quenching of long rods or plates that have been subjected to metal-working processes;
- the cyclic variation of temperature within the walls of the cylinders of an internal combustion engine;
- the heating up of a turbine rotor after starting;
- the diurnal or annual variation of temperature within the earth's crust;
- the freezing of a lake in winter.

This class of problems is also important in developing an understanding of more general heat conduction processes, because the way that space and time variations interact is more easily understood if only one space coordinate is significant. Moreover, exact analytical solutions of the differential equation governing heat conduction are more plentiful, and of simpler algebraic form, than for two- or three-dimensional situations. Thus we are able to make direct checks on the accuracy of the numerical solutions generated by the TEACH–C solution procedure. Accuracy checks form part of the objectives of LESSON 1.

The differential equation governing this class of conduction processes for the case where the heat paths are purely radial may be written:

$$\rho c_v r \frac{\partial T}{\partial t} - \frac{\partial}{\partial r}\left(kr \frac{\partial T}{\partial r}\right) - r s = 0 \tag{1.1}$$

The corresponding equation for plane one-dimensional conduction is

$$\rho c_v \frac{\partial T}{\partial t} - \frac{\partial}{\partial y}\left(k \frac{\partial T}{\partial y}\right) - s = 0 \tag{1.2}$$

In the series of computer explorations* specified below, only the case of plane heat conduction is examined (i.e. cases where the level of T is described by eqn. (1.2)). The reader may, of course, readily construct analogous problems of radial conduction by switching on the axisymmetric option in TEACH–C. In every LESSON, k, ρ and c_v are taken as invariant, though, again, the reader may explore the influence of temperature- (or space-) dependent thermal conductivity by assigning a suitable functional dependence of GAMH(I,J) in PROPS.

---

* or LESSONs as we shall call them from now on.

2. Programming: PROBLEM 1 Base Case

Instructions will now be given on how to adapt TEACH–C to the solution of equations (1.1) or (1.2) for various circumstances. The case considered is that of conduction in an infinite slab of thickness $2h$ and uniform initial temperature T_I, whose faces are suddenly heated to temperature T_B. The main points to note are firstly that the one-dimensional character of the conduction processes allows calculations to be confined to just one row of cells, and secondly that it is advantageous to select a vertical row, so that the direction of temperature variation coincides with the direction of application of the TDMA. By this choice the solution may, for linear problems, be obtained with just one application of the algorithm at each time step. The grid adopted is shown in fig. 1.1, where the symmetry of the problem has been exploited to confine calculations to the region bounded by the lower surface $y = 0$ and the symmetry plane $y = h$. The calculations are performed along the I = 2 grid-line, and the one-dimensionality condition,

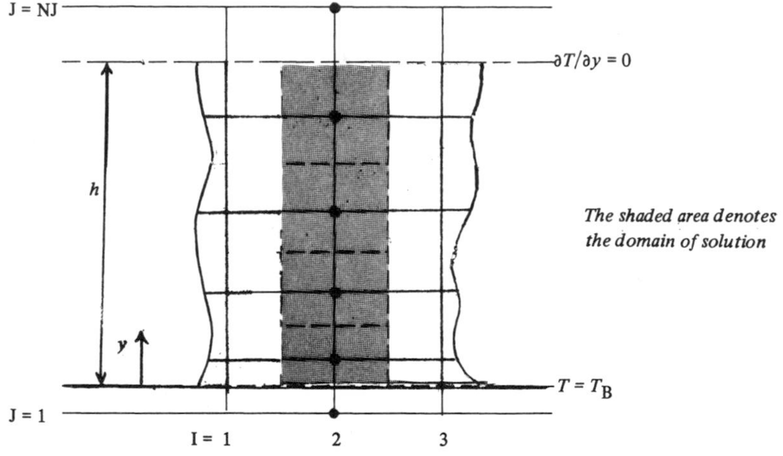

Fig. 1.1 The grid used for PROBLEM 1

which entails that there should be no heat transfer in the horizontal (x) direction, is imposed by breaking, in PROMOD, the links to the neighbouring nodes along I = 1 and I = 3. The conditions on the lower and upper boundaries of the calculation zone are also set in PROMOD. The grid is caused to expand from the lower surface, with a constant factor FEXPY between successive grid intervals.

TABLE 1.1 New FORTRAN variable for PROBLEM 1 Base Case

Variable	Quantity	Significance
TI	T_I	Initial value of temperature over whole field

TABLE 1.2 Programming instructions for PROBLEM 1 Base Case

Subroutine	Chapter	Changes
CONTRO	1	♦ Specify NI = 3, FEXPY = 1.1 and H = 0.1 ♦ Provide values of surface temperature TB and initial temperature TI ♦ Specify initial increment DT and total number of time steps MAXSTP ♦ Store initial temperature in T(2, J) ♦ Calculate residual-source normalisation factor SNORM ♦ Store steady-state solution in T(1, J) ♦ Arrange for print-out of problem-specification data ♦ Arrange for adjustment of DT and INDTIM during calculation ♦ Provide Formats for output
PROMOD	2	♦ Amend sequence for top boundary to incorporate zero-flux condition ♦ Amend side boundary sequences to incorporate zero-flux condition
PRINT	2	♦ Arrange for headings indicating contents of array columns

The remaining changes are ones of detail. They include:

- Insertion of the initial and boundary conditions and the material properties (those of steel are already specified in the TEACH–C code).
- Specification of the grid and the time intervals – here it has been arranged that DT is increased by a factor of 1.05 at each time step to enable the steady state to be reached quickly.
- Causing the temperature on the symmetry plane to be set equal to the adjacent interior value (this is a purely 'cosmetic' operation, which has no influence on the calculations).
- Arranging for the print-out of initial information and results.

These and other changes are summarised in table 1.2, above. The UPDATE listing which effects them is as follows:

PROBLEM 1 Base Case UPDATE for effecting the changes specified in Tables 1.1 and 1.2

```
*IDENT,PROBLEM1 |This is merely the name of the 'Update' – ignore it if you are updating by hand
*D,CONTRO.57    |This means remove the line labelled CONTRO 57 in TEACH–C and replace with the
      NI=3      |line that follows (NI=3)
*D,CONTRO.6Ø
      H=Ø.1
*D,CONTRO.62
      FEXFY=Ø.1
*D,CONTRO.84
      IMON=2
*D,CONTRO.92
      MAXSTP=1ØØ
*D,CONTRO.97
      SORMAX=Ø.5
*D,CONTRO.98
      DT=4.5
*D,CONTRO.111,114 This means remove lines CONTRO 111 to CONTRO 114 (inclusive) and replace
      TBOT=4ØØ.     with the two lines that follow in this listing
      TI=1ØØØ.
*D,CONTRO.115,12Ø. This means: remove lines CONTRO 115 to CONTRO 120 from TEACH–C
*I,CONTRO.122          |This means: insert immediately after line CONTRO 122 in TEACH–C
      T(2,1)=TBOT      |
      DO 221Ø J=2,NJ   |the five lines that follow (i.e. all the lines up to the next line beginning
 221Ø T(2,J)=TI        |
      DO 222Ø J=1,NJ   |with a *)
 222Ø T(4,J)=Ø.Ø
*D,CONTRO.124,127
      SNORM=ABS(TCON*(TBOT-TI)*DX/H)
C-----COMPUTE STEADY-STATE SOLUTION AND STORE AS T(1,J)
      DO 223Ø J=1,NJ
 223Ø T(1,J)=TBOT
*D,CONTRO.129,13Ø
      WRITE(6,29ØØ)H,CV,TCON,DENSIT,DT,TI,SNORM,NI,NJ
      CALL PRINT(1,1,4,NJ,IT,JT,X,Y,T,HEDT)
*I,CONTRO.136
C-----PRINTOUT LESS OFTEN AFTER FIRST FEW STEPS
      IF(NSTEP.GT.5)NSTPRI=5
*I,CONTRO.136
C-----INCREASE TIME STEP
      IF(NSTEP.GT.1)DT=1.Ø5*DT
*D,CONTRO.149
      T(2,NJ)=T(2,NJM )
*D,CONTRO.156
      CALL PRINT(1,1,4,NJ,IT,JT,X,Y,T,HEDT)
*D,CONTRO.172
      CALL PRINT(1,1,4,NJ,IT,JT,X,Y,T,HEDT)
```

```
*D,CONTRO.181
     1     CALL PRINT*114NJ,T,JT,X,Y,T,HEDT)
*D,CONTRO.184,185
 2900 FORMAT(/16X,60H PROBLEM 1      UNSTEADY ONE-DIMENSIONAL CONDUCTION P
     1ROCESSES         /
*D,CONTRO.187
*I,CONTRO.191
     2/16X,40H INITIAL TEMPERATURE OF SOLID, TI-------=,1PG10.3,2H C
*D,PROMOD.24,28
*D,PROMOD.35
*D,PROMOD.38,40
*D,PROMOD.43
*D,PROMOD.46,48
*D,PRINT.49
C-----THERE ARE NO MEANINGFUL X-VALUES FOR A 1-D PROBLEM, SO
C-----WE DO NOT PRINT THEM OUT
*D,PRINT.53
 1800 FORMAT(//1X,7(2H*-),7X,6A6,7X,7(2H-*)/
     1/7X,6HSTEADY,4X,6HCALC D
     1/7X,6HSTATE ,4X,6H TEMP )
```

3. LESSONs

3.1 LESSON 1 Plane slab with step change in surface temperature

3.1.1 *Physical background*

As in the Base Case, we consider an infinite, uniform-property slab (e.g. a steel plate) of thickness $2h$, initially at a uniform temperature T_I, as indicated in fig. 1.2. There are no heat sources present. At a time $t = 0$, the temperature of the top and bottom surfaces is abruptly changed to T_B and held fixed at that value thereafter. The equality

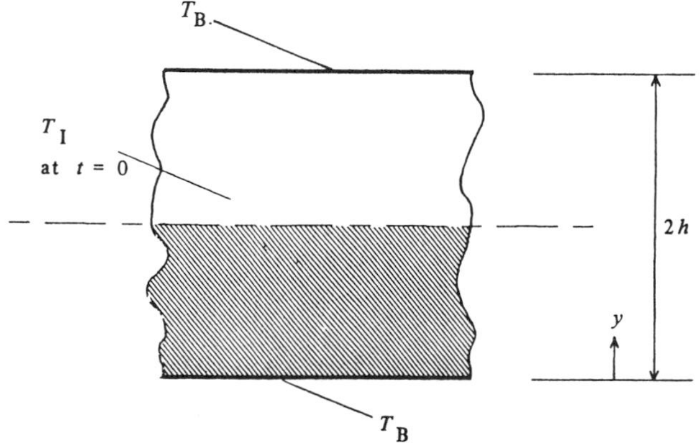

Fig. 1.2 The plane slab

of temperature at the top and bottom surfaces ensures that, as time proceeds, the temperature field within the slab develops symmetrically about the plane $y = h$. We may therefore limit attention to the region $0 \leqslant y \leqslant h$; symmetry requires that the temperature gradient be zero at all times at the plane $y = h$. A mathematical statement of the situation is:

$$\rho c_v \frac{\partial T}{\partial t} - k \frac{\partial^2 T}{\partial y^2} = 0 \qquad (1.3)$$

$$t = 0: \quad T = T_I \qquad \text{for all } y > 0$$

$$t \geqslant 0: \begin{cases} T = T_B & \text{at } y = 0 \\ \dfrac{\partial T}{\partial y} = 0 & \text{at } y = h \end{cases}$$

A complete analytical solution of this problem is possible (see for example the textbooks by Schneider [ref. 1.1] or Carslaw & Jaeger [ref. 1.2]) expressible in terms of a series. An interesting special case arises for small values of t. The variations in temperature are then, for all practical purposes, confined to small y, so that the distribution of temperature is not affected by the boundary condition at $y = h$; indeed for this initial period the space/time temperature variation is the same as that in a semi-infinite block which has a surface at $y = 0$, but which extends to infinity in the positive y direction. For this case the interior boundary condition on temperature becomes:

$$T \to T_I \text{ as } y \to \infty$$

The temperature at any point in the interior may then be shown to depend solely on a combined space-time coordinate proportional to $y(\rho c_v/kt)^{1/2}$. The dimensionless temperature may be expressed in the following simple form:

$$T^* \equiv (T - T_B)/(T_I - T_B) = \frac{2}{\sqrt{\pi}} \int_0^{\eta} e^{-\eta^2} \, d\eta \qquad (1.4)$$

and $\eta \equiv \frac{1}{2} y (\lambda t)^{-1/2}$, where $\lambda \equiv k/\rho c_v$ is the *thermal diffusivity* of the material.

The heat transfer rate at the surface may be obtained by taking the derivative of eqn. (1.4). Thus:

$$\dot{q}_B = -k \left. \frac{\partial T}{\partial y} \right|_{y=0} = \frac{2}{\sqrt{\pi}} (\rho c_v k/t)^{1/2} \, (T_I - T_B) \qquad (1.5)$$

The quantity $(\lambda t)^{1/2}$ has the dimension of a length, and it may be interpreted as a measure of the distance that conduction effects penetrate in time t. Indeed it is sometimes called the *penetration depth*. It is reasonable therefore that for the duration of semi-infinite behaviour the temperature distribution at any time should depend simply on the ratio of the physical distance y to this penetration depth – i.e. on the value of η.

Of course, for the particular situation considered here, the effects of the change in surface temperature eventually penetrate to the symmetry plane and eqn. (1.4) does not from then on provide the correct temperature distribution. Instead, the temperature at any point is now a function of two dimensionless parameters, which may be expressed as the normalised position y/h and the 'time' $\lambda t/h^2$: the latter, which is simply the ratio of the elapsed time t to the time h^2/a taken to penetrate to the centre of the slab, is sometimes called the *Fourier number* Fo.

3.1.2 *Objectives*

(*a*) To obtain numerical predictions of the temperature–time history of steel and copper slabs subjected to a step change in surface temperature, and to provide a physical explanation of the observed behaviour.

(*b*) To ascertain the influence of the grid and time increments on the accuracy of the numerical solutions.

3.1.3 *Method*

Make two RUNs, one for a steel slab and the other for a copper slab for the following conditions:

- slab half thickness h = 0.1 m,
- initial temperature T_I = 1000 K,
- surface temperature T_B = 400 K.

The number of grid nodes NJ should be specified as 12, the expansion factor FEXPY set to 1.1 and the initial time steps DT made equal to 0.045 for steel and 0.0065 for copper. It should be arranged in each case that output is generated for a value of Fourier number of 0.5. The relevant thermophysical properties of copper and steel in S.I. units are as follows:

	DENSIT (ρ) kg/m^3	CV (c_v) J/kg K	TCON (k) J/m s K
Steel	7850	460	52
Copper	8960	380	390

For either steel or copper make four further RUNs in which

◇ the initial time increment DT is first halved and then doubled, and then
◇ the value of NJ is first increased to 20 and then reduced to 6 (thereby altering the grid-spacing).

TABLE 1.3 New FORTRAN variables for LESSON 1

Variable	Quantity	Significance
DTO		Old value of DT; used to reset DT after printout at t = FOUTIM
ETA	η	Normalised y-coordinate
FOUPRI		User-assigned value of Fourier number at which printout is generated; here set to 0.5
FOURI	Fo	Fourier number
FOUTIM		Time corresponding to FOUPRI
QFLUX	$\dot{q}''$	Computed heat flux at lower boundary
QFLUXE		Exact analytical value of heat flux for a semi-infinite medium, calculated from eqn. (1.5)

TABLE 1.4 Programming instructions for LESSON 1

Subroutine	Chapter	Changes
CONTRO	1	♦ Modify print-out frequency
		♦ Assign FOUPRI and compute FOUTIM
		♦ Modify initial time step DT
	2	♦ Initialise normalised temperatures
	3	♦ Increase DT and modify print-out frequency
		♦ Adjust time step to calculate and print at t = FOUTIM
		♦ Compute and store η, T^* and Fourier number
		♦ Compute and print numerical and exact heat fluxes at lower boundary
PRINT	2	♦ Modify column headings

3.1.4 *Programming instructions*

The configuration under study in this LESSON is the same as that of the PROBLEM Base Case. The UPDATE file therefore mainly consists of additional FORTRAN instructions designed to facilitate analysis of the computed temperature distribution. The normalised temperature field T^* and η (see eqn. (1.4)) are computed and printed in the arrays normally reserved for T (4, J) and T (3, J) respectively. Heat fluxes at the lower boundary are computed and for comparison those given by the analytical solution for the semi-infinite block (eqn. (1.5)) are also printed. The value of the Fourier number is printed whenever profiles are printed; moreover, instructions are included to trigger printout at any preset value of Fourier number. The meaning of the new FORTRAN variables introduced is given in table 1.3. The changes required are listed in table 1.4. A listing of the UPDATE required to effect these changes appears below;

PROBLEM 1 LESSON 1 UPDATE for effecting the changes specified in table 1.4

```
*IDENT,P1L1
*D,CONTRO.41
     1     /'          ','TEMPER','ATURE ','(K)    ',2*'         '/
*D,CONTRO.94
      NSTPRI=25
      FOUPRI=0.5
      FOUTIM=FOUPRI*DENSIT*CV*H*H/TCON
*D,PROBLEM1.7
      DT=0.045
*D,PROBLEM1.14
 2220 T(4,J)=1.0
      T(4,1)=0.0
*D,PROBLEM1.22,24
      DT=1.10*DT
C-----PRINTOUT MORE FREQUENTLY WHEN TIME STEP IS LARGE
      IF(NSTEP.GT.25)NSTPRI=5
*I,CONTRO.137
C-----ADJUST TIME STEP TO GET FOURIER NUMBER TO FOUPRI
      IF (TIME.EQ.FOUTIM) DT=DTO
      DTO=DT
      IF (TIME.LT.FOUTIM.AND.TIME+DT.GE.FOUTIM) DT=FOUTIM-TIME
*I,CONTRO.153
C-----COMPUTE ETA AND STORE AS T(3,J)
      TERM=0.5*SQRT(DENSIT*CV/TCON)
      DO 3210 J=2,NJM1
      ETA=TERM*Y(J)/SQRT(TIME)
 3210 T(3,J)=ETA
C-----T(4,J) CONTAINS NORMALISED TEMPERATURES
      DO 3220 J=1,NJ
 3220 T(4,J)=(T(2,J)-AMIN1(TI,TBOT))/ABS(TI-TBOT+1.E-15)
C-----FOURI CONTAINS THE FOURIER NUMBER
      FOURI=TCON/DENSIT/CV/H**2*TIME
*D,CONTRO.170,171
C-----OUTPUT AS SPECIFIED BY NSTPRI OR FOR SPECIFIED FOURIER NO.
 3500 IF(MOD(NSTEP,NSTPRI).NE.0.AND.TIME.NE.FOUTIM)GOTO 3600
*I,PROBLEM1.27
C-----OBTAIN HEAT FLUX AT LOWER WALL FROM NUMERICAL AND EXACT SOLUTIONS
      QFLUX=-GAMH(2,1)*(T(2,2)-T(2,1))/Y(2)
      QFLUXE=SQRT(DENSIT*CV*TCON/(PI*TIME))*(TBOT-TI)
      WRITE(6,3901)QFLUX,QFLUXE,FOURI
*I,PROBLEM1.30
     1/16X,48H LESSON 1      STEP CHANGE IN SURFACE TEMPERATURE//
*I,CONTRO.201
 3901 FORMAT(/2X,29HHEAT FLUXES AT LOWER BOUNDARY,10X,7HFOURIER
     1/2X,29H(NUMERICAL)     (SEMI-INFINITE),10X,6HNUMBER/1H ,1PG12.3,2X,
     11PG12.3,7H W/M**2,3X,1PG12.3)
*D,PROBLEM1.35,36
     1/7X,6HSTEADY,4X,6HCALC D,4X,6H ETA  ,4X,6HNORM D
     1/7X,6HSTATE ,4X,6H TEMP ,14X,6H TEMP )
```

The following UPDATE effects changes to all the parameters involved in the RUNs of LESSON 1. Not all the changes will be needed for all the RUNs.

```
*IDENT,P1L1R1
*D,CONTRO.58
      NJ=2Ø
*D,CONTRO.87,89
      TCON=39Ø.
      CV=38Ø.
      DENSIT=896Ø.
*D,P1L1.5
      DT=.ØØ65
```

3.1.5 *Analysis of results*

(i) Plot on graph 1.1* the calculated profiles of T^* versus η for both steel and copper, confining attention to times for which the centreline value of T^* has changed by less than 1% from its initial level. Use different symbols for the different times and materials. What level of agreement is obtained with the exact solution for the semi-infinite block, given by eqn. (1.4) and shown as the solid curve on the graph?

(ii) On graph 1.2 plot the temperature profiles for both steel and copper at the larger times at which the Fourier numbers are equal to 0.5. Does the observed behaviour indicate that T^* is a universal function of y/h and Fourier number, irrespective of the material properties?

(iii) What is the approximate ratio of the times for the copper and steel slabs to reach the steady state (mid-plane temperature not more than 1 K above the surface temperature)? Is this ratio related in the expected way to the ratio of the thermal diffusivities? Explain your answers.

(iv) On graph 1.3 insert the predicted profiles of T^* versus Fourier number for $y/h = 0.5$, for the various grid- and time-intervals explored. What conclusions can be drawn concerning grid- and time-interval independence?

* Graphs will be found on pages 94–100.

3.2 LESSON 2 The effect of thermal resistance in the fluid layer

3.2.1 *Physical background*

Here the physical situation is the same as for LESSON 1 except that now we make a more general and physically realistic statement about conditions at the surface itself. We suppose that at time $t = 0$ the surface of the metal slab is exposed to a fluid of known temperature T_F and that the surface heat-transfer coefficient α between the fluid and metal is also known. The boundary condition at the surface $y = 0$ (see fig. 1.2) for times $t > 0$ thus becomes:

$$(\dot{q}''_B) = -k \left.\frac{\partial T}{\partial y}\right|_{y=0} = -\alpha\ (T_B - T_F) \qquad (1.6)$$

where in general T_B will not now remain constant as the conduction process proceeds. Moreover, even for small times, equation (1.4) is not valid except in special circumstances (see below).

The presence of α adds a further dimensionless group to the problem, $(\alpha h/k)$ which, in the present context, is often referred to as the *Biot number*, Bi. Physically, this group represents the ratio of the thermal resistance of the block to that of the fluid layer. When Bi is very large, the convection layer offers negligible resistance compared with the internal resistance of the block; under these conditions the temperature variation in the block will then be as found in LESSON 1. At the other extreme, when Bi is very small, implying that the convection layer is now the dominant resistance, the slab temperature is uniform and varies with t according to:

$$T^* \equiv \frac{T - T_F}{T_I - T_F} = \exp\left\{-\frac{\alpha t}{h\rho c_v}\right\}$$

$$= \exp(-\text{Bi.Fo}) \qquad (1.7)$$

This result, which is given by Schneider [ref. (1.1)], is easily derived by integrating eqn. (1.2), imposing the boundary condition (1.6) and assuming that the temperature T is uniform.

3.2.2 *Objective*

To explore the effects of a finite convective thermal resistance at the plate surface on the nature of the conduction processes within a steel slab suddenly immersed in a fluid of different temperature.

3.2.3 *Method*

The calculations will be made for the slab of 0.2m thickness. Firstly, incorporate the boundary condition given by equation (1.6) in the usual way, through the linearised source coefficients. Then run the program for values of the Biot number 10^2, 1 and 10^{-3} (obtained by resetting α) using the thermo-physical properties of steel.

Take the number of nodes in the y direction as 16 and specify the initial time step according to the table given below.

BIOT	DT
0.001	50.0
1.0	0.05
100.0	0.015

3.2.4 *Programming instructions*

The principal alteration to the program consists of inserting the convection boundary condition in PROMOD, altering the input and output specification, and inserting for SNORM the expression

$$|(T_F - T_{y=h})/(h/k + 1/\alpha)|$$

which is a notional heat flux based on the thermal resistance concept. These and other changes are summarised in table 1.6, which is followed by the LESSON and RUN UPDATE listings.

TABLE 1.5 New **FORTRAN** variables for **LESSON 2**

Variable	Quantity	Significance
ALPHA	α	Surface heat transfer coefficient
BIOT	Bi	Biot number
DTO		Old value of DT, used to reset DT following printout at FOUTIM
FOUPRI		User-assigned value of Fo for which printout is required
FOURI	Fo	Fourier number
FOUTIM		Time corresponding to FOUPRI
TF	T_F	Fixed fluid temperature

TABLE 1.6 Programming instructions for LESSON 2

Subroutine	Chapter	Change required
CONTRO	0	♦ Insert ALPHA and TF in COMMON
	1	♦ Modify printout frequency and DT
		♦ Add FOUPRI and find FOUTIM
	2	♦ Set ALPHA, TF, TI and compute Biot number, BIOT
		♦ Initialise normalised temperatures
		♦ Modify SNORM calculation
		♦ Modify steady-state solution
		♦ Print out data for LESSON
	3	♦ Increase DT and modify printout frequency
		♦ Adjust time-step to calculate at specified Fourier number, FOUPRI
		♦ Compute surface temperature
		♦ Compute normalised temperatures and Fourier number
		♦ Print Fourier number and ensure printout at specified Fourier number
	4	♦ Provide formats for LESSON
PROMOD	0	♦ Insert ALPHA and TF in COMMON
	1	♦ Insert convection boundary condition at lower boundary
PRINT	2	♦ Modify column headings

PROBLEM 1 LESSON 2 UPDATE for effecting the changes specified in Table 1.6

```
*IDENT,P1L2
*I,CONTRO.37
     1/P1L2A/ALPHA,TF
*D,CONTRO.41
     1     /'       ','TEMPER','ATURE ','(K)    ',2*'       '/
*D,PROBLEM1.5
      MAXSTP=250
*D,CONTRO.94
      NSTPRI=25
      FOUPRI=0.5
      FOUTIM=FOUPRI*DENSIT*CV*H*H/TCON
*D,PROBLEM1.7
      DT=0.05
*D,PROBLEM1.8
C-----FLUID TEMPERATURE AND HEAT TRANSFER COEFFICIENT
      TF=400.0
      ALPHA=520.
      BIOT=ALPHA*H/TCON
      RATT=TCON/ALPHA/Y(2)
      TBOT=1000.
*D,PROBLEM1.14
 2220 T(4,J)=(T(2,J)-TF)/(TI-TF+1.E-15)
*D,PROBLEM1.15
      SNORM=ABS((TF-T(2,NJ))/(H/TCON+1./ALPHA))
*D,PROBLEM1.18
 2230 T(1,J)=TF
```

```
*D,PROBLEM1.19
      WRITE(6,2900)H,CV,TCON,DENSIT,DT,TI,TF,ALPHA,BIOT,SNORM,NI,NJ
*D,PROBLEM1.22,24
      IF (NSTEP.LE.120) DT=1.06*DT
*I,CONTRO.137
C-----ADJUST TIME STEP TO SET FOURIER NUMBER TO FOUPRI
      IF (TIME.EQ.FOUTIM) DT=DTO
      DTO=DT
      IF (TIME.LT.FOUTIM.AND.TIME+DT.GE.FOUTIM) DT=FOUTIM-TIME
*I,CONTRO.148
      SNORM=ABS((TF-T(2,NJ))/(H/TCON+1./ALPHA))
      T(2,1)=(RATT*T(2,2)+TF)/(RATT+1.)
*I,CONTRO.153
C-----T(4,J) CONTAINS NORMALISED TEMPERATURES
      DO 3220 J=1,NJ
 3220 T(4,J)=(T(2,J)-TF)/(TI-TF+1.E-15)
C-----FOURI CONTAINS THE FOURIER NUMBER
      FOURI=TCON/DENSIT/CV/H**2*TIME
*D,CONTRO.170,171
C-----OUTPUT AS SPECIFIED BY NSTPRI OR FOR SPECIFIED FOURIER NO.
 3500 IF(MOD(NSTEP,NSTPRI).NE.0.AND.TIME.NE.FOUTIM)GOTO 3600
*I,PROBLEM1.27
      WRITE (6,3901) FOURI
*I,PROBLEM1.30
      1/16X,50H LESSON 2      PRESCRIBED HEAT TRANSFER COEFFICIENT//
*I,PROBLEM1.31
      2/16X,40H FLUID TEMPERATURE, TF ----------------=,G10.3,2H K
      2/16X,40H HEAT TRANSFER COEFFICIENT, ALPHA -----=,G10.3,9H W/M**2 K
      2/16X,40H BIOT NUMBER, BIOT --------------------=,G10.3
*I,CONTRO.201
 3901 FORMAT (/17H FOURIER NUMBER =,1PG12.3)
*I,PROMOD.15
      1/P1L2A/ALPHA,TF
*D,PROMOD.31,32
      RT=1./(ALPHA*SEW(IL))+1./DS
      SU(2)=SU(2)+TF/RT
      SP(2)=SP(2)-1./RT
*D,PROBLEM1.35,36
     1/7X,6HSTEADY,4X,6HCALC D,14X,6HNORM D
     1/7X,6HSTATE ,4X,6H TEMP ,14X,6H TEMP )
```

The following UPDATE effects changes to all the parameters involved in the RUNs of LESSON 2. Not all the changes will be needed for all the RUNs.

```
*IDENT,P1L2R1
*D,CONTRO.58
         NJ=16
*D,P1L2.7
         DT=5Ø.
*D,P1L2.1Ø
         ALPHA=.52
```

3.2.5 *Analysis of results*

(*i*) Examine the effect of Biot number on the temperature distribution by plotting T^* against y/h for Bi = 10^2; 1 and 10^{-3} in graphs* 1.4, 1.5 and 1.6 respectively. Show profiles over the entire time-span to the steady state, including in each case the result for Fo = 0.5. Insert on each graph the approximate time required to reach steady state. Explain the results, paying special attention to the latter feature, the shape of the temperature profiles and the behaviour of the surface temperature. Are the limiting cases of §3.2.1 realised?

(*ii*) Plot on graph 1.7 the variation of the dimensionless temperature difference

$$|(T_B - T_{y=h}) \;/\; (T_B - T_F)|$$

with Bi, for a value of Fo = 0.5. Is the behaviour consistent with the interpretation of Bi as a ratio of thermal resistances?

(*iii*) Plot in the semi-logarithmic coordinates of graph 1.8 the variation of T^* with Fo for Bi = .001. Do the results agree with the analytical solution eqn. (1.7))?

* **Graphs will be found on pages 26 onwards**

3.3 LESSON 3 Conduction with a temperature-dependent heat source

3.3.1 *Physical background*

There are numerous industrial problems of heat conduction involving heat generation within the conductor. Examples are: the heat generation in the walls of a rolling tyre, the heat release in the fuel elements of a nuclear reactor, and the heating of an electrical resistor. In the last case, the resistivity of the material of which the conductor is made will sometimes be a function of its temperature, and in some circumstances the heat-generation rate will consequently be a linear function of the temperature. It is therefore a dependence of this type that we shall consider in the present LESSON, viz.

$$s = \dot{q}_0''' \left[1 + \beta(T - T_0)\right] \tag{1.7}$$

where $\dot{q}_0'''$ is the value of the volumetric heat source at a reference temperature T_0, and β is the coefficient of dependence of resistivity on temperature.

The form of equation (1.7) immediately suggests the possibility of runaway temperatures in the conductor. If β is positive, the magnitude of the heat source will increase as the temperature rises — an inherently unstable situation. It is likely therefore that, for all values of β greater than some value to be determined, heat will not be conducted away fast enough to permit the establishment of a steady-state temperature distribution.

In this LESSON we explore this expectation by examining the electrical conductor strip sketched in fig. 1.3. We suppose that at time $t = 0$ the current flowing through the conductor (which is then at the ambient temperature T_F) is suddenly increased to a large value and held there.

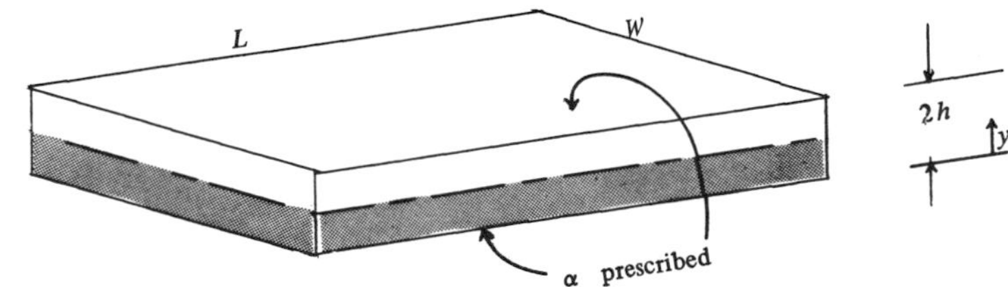

Fig. 1.3 Sketch of the electrical conducting strip

The length and width of the strip (L and W) are both presumed very much greater than its thickness $2h$. The temperature variation within the strip may thus be assumed to occur only in the y-direction. The boundary conditions on the top and bottom surfaces ($y=0$ and $y=2h$) are the same as for LESSON 2, namely a prescribed and uniform heat-transfer coefficient. The surface $y = h$ is a symmetry plane, so that computations need only be performed over the half thickness of the strip in the manner of LESSONs 1 and 2 (see figure 1.3). Dimensional analysis suggests that (if we arbitrarily set $T_0 = T_F$) the temperature distribution across the plate is dependent on four dimensionless variables, viz:

$$\beta(T-T_F) = f(y/h, \text{Fo}, \text{Bi}, S) \tag{1.8}$$

where $S \equiv (\dot{q}_0'''\beta h)/(k/h)$ is a dimensionless heat source which can be viewed as the ratio of the additional heat generation to the additional heat conduction consequent on a unit rise in the temperature of the strip. An analytical solution of eqn. (1.8) exists for the case of very large time (steady-state condition) and very large Biot number. The solution is [ref. (1.4)]:

$$T^* \equiv \beta(T - T_F) = \frac{\cos\dfrac{(h-y)\sqrt{S}}{h}}{\cos\sqrt{S}} - 1 \tag{1.9}$$

We see that for values of $\cos\sqrt{S}$ near zero, the strip temperature must become very large. The critical condition where $\cos\sqrt{S}$ is identically zero occurs when $S = \pi^2/4$. Failure of the conductor is then inevitable, regardless of the material of which it is composed.

As mentioned, eqn (1.9) applies when the Biot number is very large. As the Biot number is steadily reduced, implying greater resistance to heat transfer between the surface and the surrounding fluid, the critical value of S must be expected to diminish. The fact that materials for which β is positive exhibit critical values of S is the basis of the behaviour of the common electrical fuse.

3.3.2 *Objective*

To explore the character of time-dependent heat conduction in a strip of conducting medium for a heat source which is linearly dependent on temperature.

3.3.3 *Method*

We choose to examine the behaviour of a strip of copper which, in common with most pure metals, has a positive temperature coefficient of resistance ($\beta = 4 \times 10^{-3}\ \text{K}^{-1}$). We do this by way of the following two sets of RUNs:

(*i*) Adjust α so that Bi is large (about 10^3 say) and perform three RUNs for separate values of $\dot{q}_0'''$, such that S is equal to 1, 2, and 3

(*ii*) For $S = 1$, decrease Bi successively to 102.6, 10.26, 1.026. and 0.1026 by varying α.

3.3.4 *Programming instructions*

TABLE 1.7 New FORTRAN variables for LESSON 3

Variable	Quantity	Significance
ALPHA	α	Surface heat transfer coefficient
BETA	β	Temperature coefficient of resistivity
BIOT	Bi	Biot number
Q0	$\dot{q}_0'''$	Heat-source at reference temperature
S	S	Non-dimensional source-strength parameter
T0	T_0	Reference temperature
TF	T_F	Fixed fluid temperature

TABLE 1.8 Programming instructions for LESSON 3

Subroutine	Chapter	Change required
CONTRO	0	♦ COMMON block to include ALPHA, BETA, Q0 T0 and TF
	1	♦ Height of calculation zone = .0001m
		♦ Incorporate material properties of copper
		♦ Set boundary values
		♦ Set DT
		♦ Specify source-strength parameters and the heat-transfer coefficient ALPHA
	2	♦ Print out LESSON heading and data
		♦ Calculate Biot Number and S
	3	♦ Calculate correctly normalised temperatures
	4	♦ Modify formats for output as required
PROMOD	0	♦ Arrange for COMMON block to correspond to CONTRO
	1	♦ Incorporate temperature-dependent heat source and convective boundary condition at lower boundary
PRINT	2	♦ Modify column headings

The new FORTRAN variables for LESSON 3 are listed in table 1.7. The changes required to produce the LESSON UPDATE from the PROBLEM Base Case are straightforward: they are summarised in table 1.8.

PROBLEM 1 LESSON 3 UPDATE for effecting the changes specified in table 1.8

```
*IDENT.P1L3
*I.CONTRO.39
      1/P1L3A/QØ.BETA.TØ.ALPHA.TF
*D.CONTRO.43
      1      /6H       .6HTEMPER.6HATURE .6H(K)    .2*6H
*D.PROBLEM1.2
       H=1.E-Ø4
*D.CONTRO.66.69
C-----MATERIAL PROPERTIES OF COPPER
       TCON=39Ø.
       CV=38Ø.
       DENSIT=896Ø.
*D.PROBLEM1.5
       MAXSTP=95
*I.PROBLEM1.6
       TI=3ØØ.
       TBOT=3ØØ.
*D.PROBLEM1.7
C-----CALCULATE AN APPROPRIATE VALUE OF DT FOR THE LESSON
       DT=CV*H*H*DENSIT/TCON*3./226.
C-----SOURCE STRENGTH PARAMETERS
       QØ=1.ØE+13
       BETA=4.ØE-Ø3
       TØ=TI
C-----FLUID TEMPERATURE AND HEAT TRANSFER COEFFICIENT
       TF=3ØØ.
       ALPHA=4.ØE+9
*D.PROBLEM1.6.9
*D.PROBLEM1.15
       SNORM=QØ*H*DX
*D.PROBLEM1.19
       S=QØ*(H**2)*BETA/TCON
       BIOT=ALPHA*H/TCON
       WRITE(6,29ØØ)H.CV.TCON.DENSIT.DT.TI.TF.ALPHA.BIOT.QØ.BETA.TØ.S.
      1      SNORM.NI.NJ
*I.CONTRO.153
C-----T(4.J) CONTAINS NORMALISED TEMPERATURES
       DO 322Ø J=1.NJ
 322Ø T(4,J)=(T(2,J)-T(2.1))/(T(2.NJM1)-T(2.1)+1.E-15)
*I.PROBLEM1.3Ø
      1/16X.47H LESSON 3     CONVECTIVE BOUNDARY CONDITION AND
      1/3ØX.33HTEMPERATURE-DEPENDENT HEAT SOURCE//
*I.PROBLEM1.31
      2/16X.4ØH FLUID TEMPERATURE. TF ----------------=.G1Ø.3.2H K
      2/16X.4ØH HEAT TRANSFER COEFFICIENT, ALPHA -----=.G1Ø.3.9H W/M**2 K
      2/16X.4ØH BIOT NUMBER. BIOT --------------------=.G1Ø.3
      2/16X.4ØH SOURCE PARAMETERS. QØ ----------------=.G1Ø.3.7H W/M**3
      2/16X.4ØH ..................., BETA -------------=.G1Ø.3.6H K**-1
      2/16X.4ØH ..................., TØ ---------------=.G1Ø.3.2H K
      2/16X.4ØH ..................., S ----------------=.G1Ø.3
*I.PROMOD.15
      1/P1L3A/QØ.BETA.TØ.ALPHA.TF
*I.PROMOD.18
       DO 1Ø1Ø J=2.NJM1
       DV=RY(J)*SEW(IL)*SNS(J)
       SU(J)=SU(J)+QØ*(1.-BETA*TØ)*DV
 1Ø1Ø SP(J)=SP(J)+QØ*BETA*DV
*D.PROMOD.31.32
       RT=1./(ALPHA*SEW(IL))+1./DS
       SU(2)=SU(2)+TF/RT
       SP(2)=SP(2)-1./RT
*D.PROBLEM1.35.36
      1/7X.6HSTEADY.4X.6HCALC D.14X.6HNORM D
      1/7X.6HSTATE .4X.6H TEMP .14X.6H TEMP )
```

The following UPDATE effects changes to all the parameters involved in the RUNs of LESSON 3. Not all the changes will be needed for all the RUNs.

```
*IDENT,P1L3R1
*D,P1L3.14
       QØ=2.ØE13
*D,P1L3.19
       ALPHA=1.ØE8
```

3.3.5 *Analysis of results*

(*i*) Plot on graph* 1.9 the profile of the steady-state temperature, as characterised by $\beta(T-T_F)$, versus normalised position in the conductor, for $S \cong 1$ and for a Biot number of 1026. Compare the results with the analytic profile given by eqn. (1.9); this is already plotted on graph 1.9.

(*ii*) On graph 1.10 plot the maximum temperature in the conductor against time, for a Biot number of 1026 and for each of the three values of S. Discuss the conditions for which the conductor would melt (the melting temperature of copper of 1356 K is indicated on the graph). What distinguishes the value of β for a material which would be totally unsuitable for a fuse? Use a handbook to provide an example of such a material.

(*iii*) For $S = 1$ plot on graph 1.11 the value of the maximum temperature within the conductor versus time, with Bi as a parameter. What is the influence of Biot number on the critical value of S which is revealed by these results? Is this influence in line with our expectations?

(*iv*) Examine the temperature distribution within the strip at various times for Bi = .1026 and explain. Does the shape of the profiles suggest a possible simplification to the analysis?

3.4 LESSON 4 Heat conduction with sinusoidally varying surface temperature

3.4.1 *Physical background*

Consider a semi-infinite medium in which a periodic variation of temperature is applied at the surface $y = 0$. This idealisation may be † used to model the heat transfer by conduction in the earth's crust due to the diurnal or annual variation of surface temperature. The governing differential equation is the same as for LESSONs 1 and 2, i.e. eqn. (1.3). It is instructive to start the computations with a uniform initial temperature equal to the mean value, although this is not typical of conduction in the earth's crust. The solutions will thus display two distinct contributions: a "periodic" behaviour which is cyclic (with period $2\pi/\omega$) and a "transient" part which becomes negligible at large times. The boundary and initial conditions now become

$$\begin{aligned} t = 0: &\quad T = T_I && \text{for all} \quad y \geq 0 \\ t > 0: &\quad T = T_I + \Delta T \sin \omega t && y = 0 \\ &\quad T \to T_I && \text{as} \quad y \to \infty \end{aligned} \tag{1.10}$$

* Graphs will be found on pages 26 onwards

† As will be demonstrated in this LESSON, the temperature oscillations about the mean value become negligible at a finite depth below the surface.

where ΔT is the amplitude $(T_{max} - T_{ave})$ of the surface temperature variation about its mean value. The two characterising parameters of the problem are a dimensionless length scale y^* (defined as $y \{\rho c_v \omega / k\}^{1/2}$) and a time scale ωt. The analytical treatment of this class of problem is discussed by Schneider [ref. 1.1] and Schlichting [ref. 1.3]. The latter gives a complete solution for both the transient and periodic parts† in the form of an infinite trigonometric series. The periodic solution has the simple form:

$$\frac{T - T_I}{\Delta T} = e^{-y^*/\sqrt{2}} \sin(\omega t - y^*/\sqrt{2}) \tag{1.11}$$

Thus, the amplitude of the harmonic temperature variations decays exponentially with distance. The temperature of material at depth y has a phase lag $y^*/\sqrt{2}$ with respect to the surface temperature.

3.4.2 *Objectives*

To discover:

(*a*) the depths to which diurnal and annual temperature variations penetrate the earth and moon surfaces.

(*b*) the dependence of the depth of penetration on the thermal diffusivity of the medium and on the period of the temperature cycle.

(*c*) the number of cycles needed for the temperature variation to become periodic.

3.4.3 *Method*

In principle the temperature variations caused by the sinusoidally varying surface temperature extend to infinity. Common sense – and equation (1.11) – suggest, however, that temperature fluctuations about the mean value become negligible beyond a certain depth, the value of which depends on the magnitudes of ω and $(k/\rho c_v)$. In the calculations, therefore, the boundary condition at infinity is replaced by the requirement that the temperature gradient be zero at some finite depth h, which has been determined by trial and error (with some guidance from eqn (1.11), to be below the region of significant oscillations. The sinusoidal temperature variation indicated by eqn (1.11) is applied at the surface $y = 0$.

The RUNs which are to be performed are summarised in table 1.9. At each time step arrangements are made in the calculations to determine the penetration depth, which is defined as the location below the surface at which the temperature variation over a cycle is 2% of the variation at the surface: if this location does not correspond to a grid node, linear interpolation is used.

3.4.4 *Programming instructions*

Table 1.11 summarises how PROBLEM 1 may be adapted to facilitate the explorations indicated above. This is followed by listings of the LESSON 4 UPDATE and a typical RUN UPDATE.

---

† For the mathematically equivalent problem of an infinite plate oscillating in its own plane in an expanse of viscous fluid initially at rest.

TABLE 1.9 Physical parameters for RUNs of LESSON 4

RUN	*Medium*	k J /msK	ρ kg/m³	c_v J /kg/K	*Period*	T_{max} (°C)	T_{ave} (°C)
1	Earth soil	0.57	1600	2100	1 day	16	10
2	Earth soil	0.57	1600	2100	1 year	20	10
3	Moon dust [Ref. 1.5]	0.002	1300	700	1 moon day (27.2 earth days)	100	–25
4	Earth soil	0.57	1600	2100	1 hour	16	10
5	–	0.057	1600	2100	1 day	16	10
6	–	5.7	1600	2100	1 day	16	10

TABLE 1.10 New FORTRAN variables for LESSON 4

Variable	Quantity	Significance
ALPHA	α	Surface heat transfer coefficient
ANAME		Contains name of soil-type; used for heading of output
CYCLE		Elapsed time in cycles
OMEGA	ω	2π / PERIOD
PERIOD		Period of temperature fluctuations
QFLUX	$\dot{q}'''$	Heat flux at lower boundary
QFLUXE		Exact analytical heat flux for a semi-infinite medium
TAVE	T_{ave}	Average temperature of surface
TMAX	T_{max}	Maximum surface temperature
YNORM	$y☆$	Normalised depth

TABLE 1.11 Programming instructions for LESSON 4

Subroutine	Chapter	Changes
CONTRO	0	♦ Set ANAME to identify soil type
	1	♦ Assign physical properties and PERIOD. Calculate OMEGA..
		♦ Set NJ = 16 and calculate H by reference to dimensionless depth y^*.
		♦ Enlarge grid expansion factor FEXPY.
		♦ Re-assign MAXSTP, DT and printout frequency
	2	♦ Assign TMAX, TAVE and SNORM
		♦ Calculate exact periodic solution from eqn (1.11) and put in T(1,J)
		♦ Printout data for LESSON
	3	♦ Modify printout frequency
		♦ Introduce sinusoidal variation of surface temperature
		♦ Compute and store YNORM in T(3, J) and normalised temperature in T(4, J)
		♦ Delete printout of residual sources (to save lines of printout)
		♦ Calculate exact periodic solution. Compute and print numerical and exact cyclic heat fluxes at lower boundary
	4	♦ Provide formats for LESSON
PRINT	2	♦ Modify column headings

PROBLEM 1 LESSON 4 UPDATE for effecting the changes specified in table 1.11

```
*IDENT,P1L4
*D,CONTRO.39
      CHARACTER*6 HEDT(6),HEDS(6),ANAME(2)
*I,CONTRO.43
      DATA ANAME/' EARTH',' SOIL '/
*I,CONTRO.54
C-----INSERT PROPERTIES AT THIS POINT TO ENABLE HEIGHT CALCULATION
      TCON=0.57
      DENSIT=1600.
      CV=2093.
C-----SET PERIOD IN EARTH DAYS
      PERIOD=1.0
      OMEGA=PI/12./3600./PERIOD
*D,CONTRO.58
      NJ=16
*D,PROBLEM1.2
      H=10.0*(TCON/PI*3600.*24.*PERIOD/CV/DENSIT)**0.5
*D,PROBLEM1.3
      FEXPY=1.25
*D,CONTRO.85
      JMON=8
```

```
 *D,CONTRO.86,89
 *D,PROBLEM1.5
       MAXSTP=160
 *D,CONTRO.94
       NSTPRI=4
 *D,PROBLEM1.7
       DT=1./40.*24.*3600.*PERIOD
 *D,PROBLEM1.8,9
       TAVE=10.
       TI=TAVE
       TBOT=TI
       TMAX=16.
 *D,PROBLEM1.15
       SNORM=TCON*(TMAX-TI)*DX/H
 *D,CONTRO.128
       IF(NSTEP.GT.10.AND.MOD(NSTEP,NSTPRI).NE.0)GO TO 4100
 C-----CALCULATE THE EXACT PERIODIC SOLUTION AND PUT IN T(1,J)
       ALPHA=TCON/DENSIT/CV
       DO 2240 J=1,NJ
       T(1,J)=EXP(-SQRT(OMEGA/2./ALPHA)*Y(J))*
      1 SIN(OMEGA*TIME-SQRT(OMEGA/2./ALPHA)*Y(J))
       IF(J.EQ.1)T(1,J)=SIN(OMEGA*TIME)
       T(1,J)=T(1,J)*(TMAX-TAVE)+TAVE
  2240 CONTINUE
 *D,PROBLEM1.19
       WRITE(6,2900)ANAME,CV,TCON,DENSIT,DT,PERIOD,NI,NJ
 *D,CONTRO.134
 *D,PROBLEM1.21,24
 *I,CONTRO.138
       IF (NSTEP.EQ.40) NSTPRI=10
 C-----INTRODUCE SINUSOIDAL VARIATION OF TEMPERATURE
       T(2,1)=TAVE+(TMAX-TAVE)*(SIN(PI/12./3600./PERIOD*TIME))
 *D,CONTRO.152,153
 C-----COMPUTE YNORM AND STORE AS T(3,J)
       DO 3210 J=2,NJM1
       YNORM=Y(J)*SQRT(CV*DENSIT/TCON/24./3600./PERIOD)
  3210 T(3,J)=YNORM
 C-----T(4,J) CONTAINS NORMALISED TEMPERATURES
       DO 3220 J=1,NJ
 3220 T(4,J)=(T(2,J)-TAVE)/(TMAX-TAVE)
*D,CONTRO.157
*I,CONTRO.171
      DO 3510 J=1,NJ
      T(1,J)=EXP(-SQRT(OMEGA/2./ALPHA)*Y(J))*
     1 SIN(OMEGA*TIME-SQRT(OMEGA/2./ALPHA)*Y(J))
      IF(J.EQ.1)T(1,J)=SIN(OMEGA*TIME)
      T(1,J)=T(1,J)*(TMAX-TAVE)+TAVE
 3510 CONTINUE
*D,CONTRO.173
C-----OBTAIN HEAT FLUX AT LOWER WALL FROM NUMERICAL AND EXACT SOLUTIONS
      QFLUX=-GAMH(2,1)*(T(2,2)-T(2,1))/Y(2)
      QFLUXE=GAMH(2,1)*(TMAX-TAVE)*SQRT(OMEGA*0.5/ALPHA)
     1        *(COS(OMEGA*TIME)+SIN(OMEGA*TIME))
      CYCLE=TIME/PERIOD/24./3600.
      WRITE(6,3901)QFLUX,QFLUXE,CYCLE
*D,CONTRO.186
     1/16X,56H LESSON 4      PENETRATION OF CYCLIC SURFACE TEMPERATURE
     1/30X,29HPHYSICAL PROPERTIES RELATE TO ,2A6//
*D,PROBLEM1.31
     2/16X,40H PERIOD OF TEMPERATURE CYCLE, PERIOD --=,F6.2,13H    EARTH
     1DAYS
*D,CONTRO.192
*I,CONTRO.201
 3901 FORMAT(/2X,29HHEAT FLUXES AT LOWER BOUNDARY,10X,7HELAPSED
     1/2X,29H(NUMERICAL)    (EXACT CYCLIC) ,10X,6H TIME /1H ,1PG12.3,2X,
     11PG12.3,7H W/M**2,3X,1PG12.3,7HPERIODS)
*D,PROBLEM1.35,36
     1/7X,6HCYCLIC,4X,6HCALC D,4X,6HYNORM ,4X,6HNORM D
     1/7X,6HSTATE ,4X,6H TEMP ,14X,6H TEMP )
```

The following UPDATE effects changes to all the parameters involved in the RUNs of LESSON 4. Not all the changes will be needed for all the RUNs.

```
*IDENT,P1L4R1
*D,P1L4.2
      DATA ANAME/6H EARTH,6H SOIL /
*D,P1L4.4,6
      TCON=Ø.57
      DENSIT=16ØØ.
      CV=2Ø93.
*D,P1L4.8
      PERIOD=365.25
*D,P1L4.17
      TAVE=1Ø.
*D,P1L4.2Ø
      TMAX=2Ø.
```

3.4.5 *Analysis of results*

(*i*) For *one* of the RUNs 1–3 plot the distribution of normalised temperature versus depth on graph* 1.12, after 0.1, 0.2, 0.5, 1.25, 2.25, 3.25 cycles using different symbols for each plot, and explain the behaviour. How many cycles are needed for the distribution to become periodic, as indicated by the comparison of the last three plots?

(*ii*) Plot, for one RUN, the predicted variation of temperature with depth in graph 1.13 at various stages in a periodic cycle; then estimate the penetration depth from this and compare with the value printed out, to ensure that you understand – and agree with – the latter. When this has been completed enter the penetration depths and corresponding values of $y^{\star}$ for *all* RUNs in table 1.12.

TABLE 1.12 Penetration depths for RUN conditions

RUN	Penetration Depth y (m)	Dimensionless Penetration Depth $y^{\star}$
1		
2		
3		
4		
5		
6		

(*iii*) For RUNs 1, 2 and 4 plot on graph 1.14 the variation of penetration depth versus period. Is a power law dependence indicated? If so, what is the magnitude of the exponent?

* Graphs will be found on pages 26 onwards

(*iv*) For the three RUNs made with a period of one day, and with the density and specific heat for earth clay soil, plot on graph 1.15 the variation of penetration depth versus thermal conductivity. Is a power law dependence indicated? If so what is the magnitude of the exponent?

(*v*) Is the behaviour indicated in table 1.12 and in [illegible] and [illegible] consistent with the temperature profile at a given value of t being a function only of $y*$?

(*vi*) Make a few spot checks of the agreement between the computed results for the periodic state and the analytical solution.

4. References

1.1 P.J.Schneider, *Conduction Heat Transfer,* Addison Wesley, 1959.

1.2 H.S.Carslaw & J.C.Jaeger, *Conduction of Heat in Solids,* Oxford University Press, 1957.

1.3 H.Schlichting, *Boundary Layer Theory,* McGraw Hill, 1960.

1.4 B.Gebhart, *Heat Transfer,* McGraw Hill, 1971.

1.5 A.E.Wechsler, P.E.Glaser & J.A.Fountain, *Thermal Properties of Granulated Materials,* in Progress in Astronautics & Aeronautics, **28,** 1974.

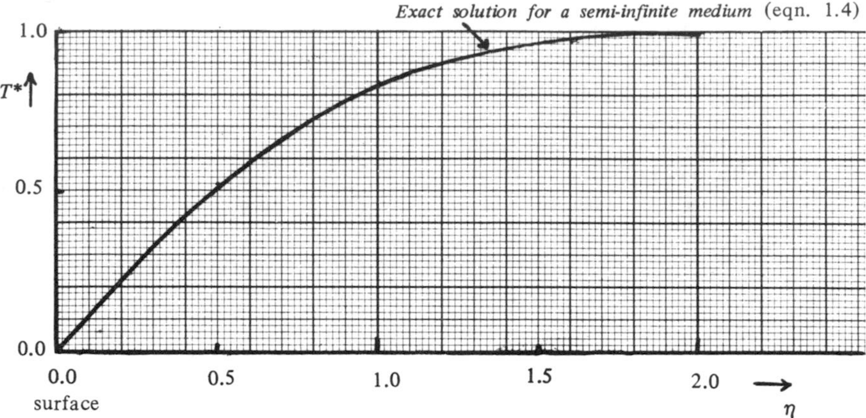

Graph 1.1 The computed profiles of T^* versus η together with the analytical solution for a semi-infinite medium.

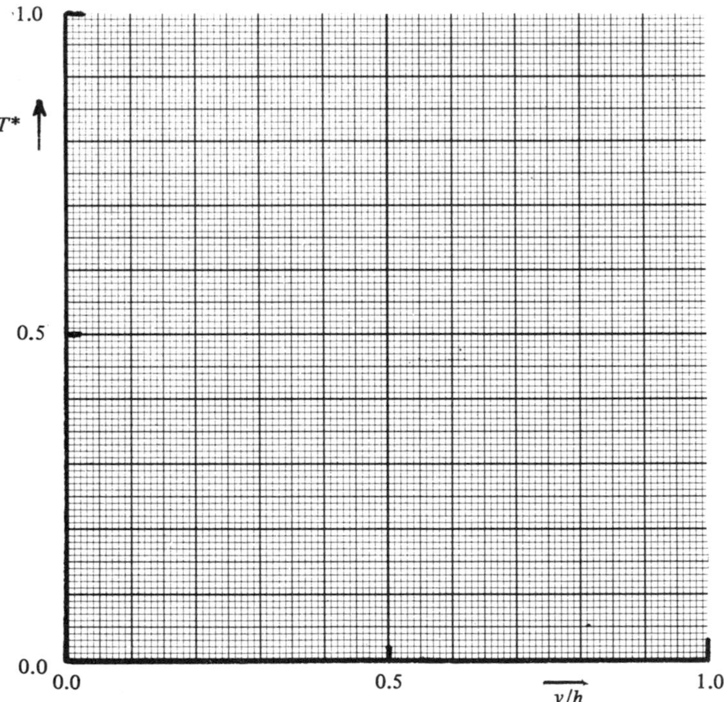

Graph 1.2 Comparison of temperature profiles for different materials at the same Fourier number

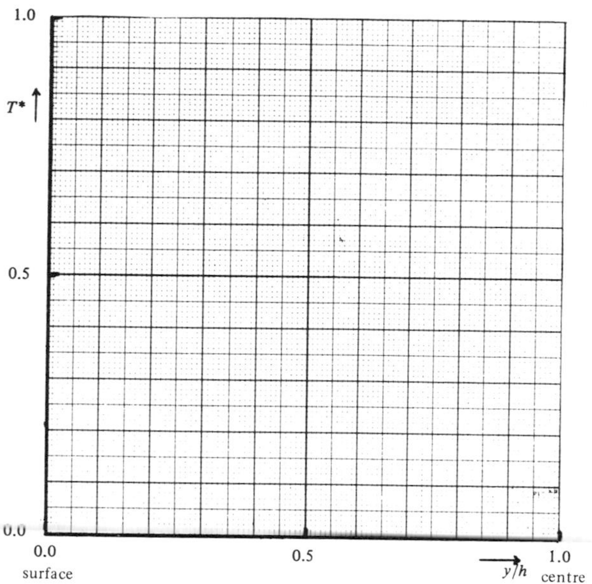

Graph 1.3 The computed profiles of T^* versus y/h for various grid and time intervals

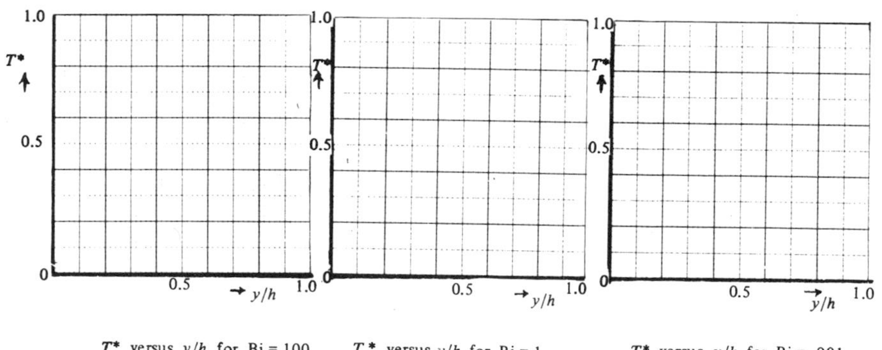

T^* versus y/h for Bi = 100

T^* versus y/h for Bi = 1

T^* versus y/h for Bi = .001

Graph 1.4

Graph 1.5

Graph 1.6

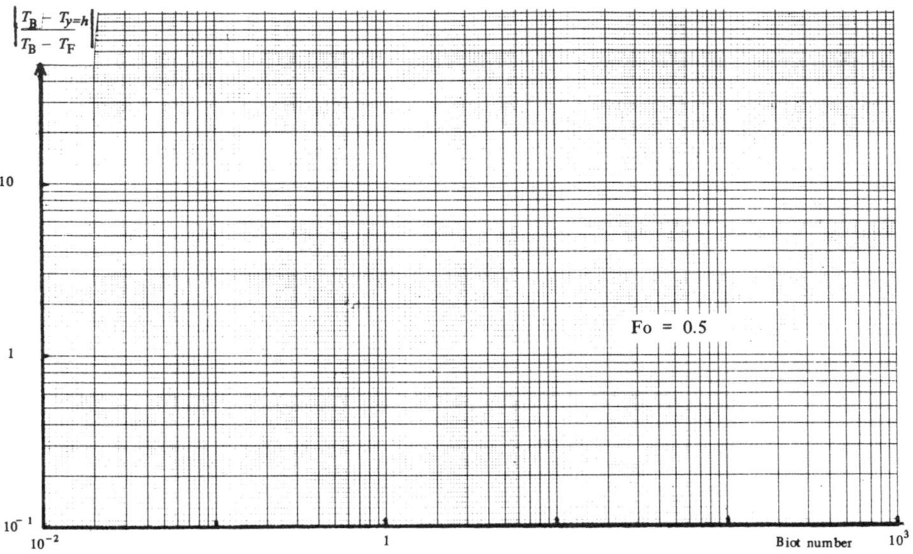

Graph 1.7 Variation of dimensionless temperature difference with Biot number for Fo = 0.5

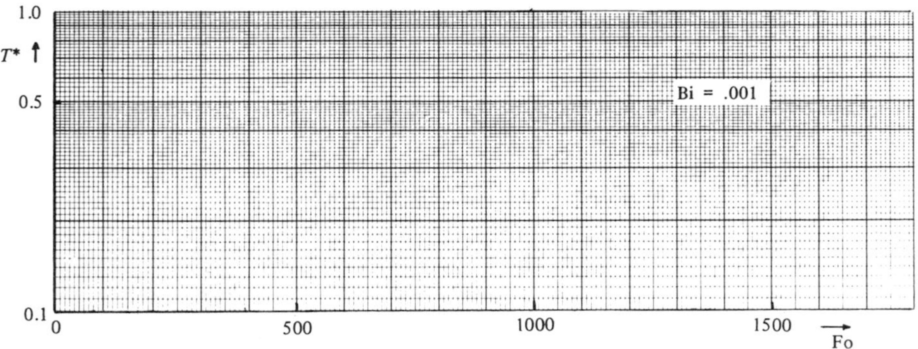

Graph 1.8 The variation of T^* with Fo for Bi = .001: comparison of numerical predictions with analytical solution given by eqn 1.7.

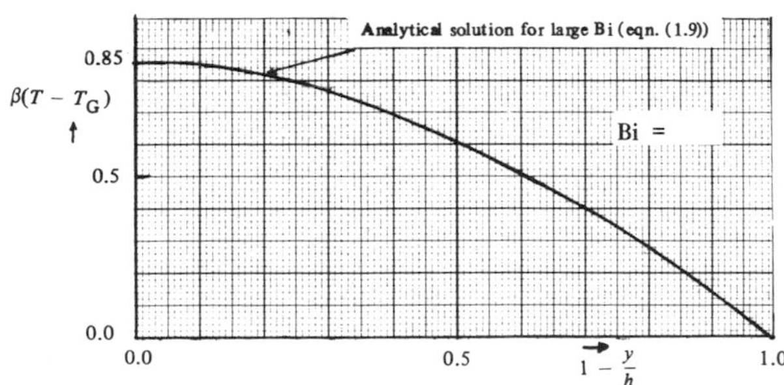

Graph 1.9 Comparison of steady-state computed and analytical solutions

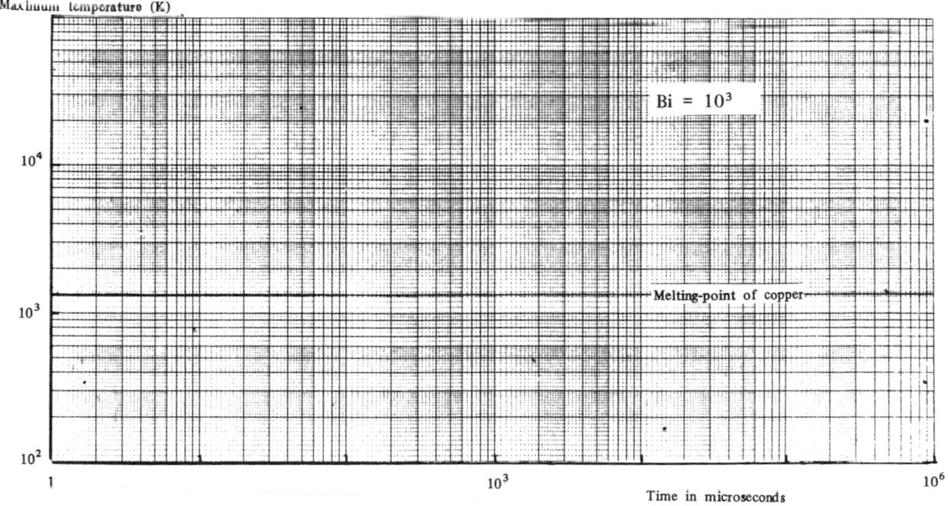

Graph 1.10 Maximum temperature in the conductor against time, various S, for Bi = 1000.

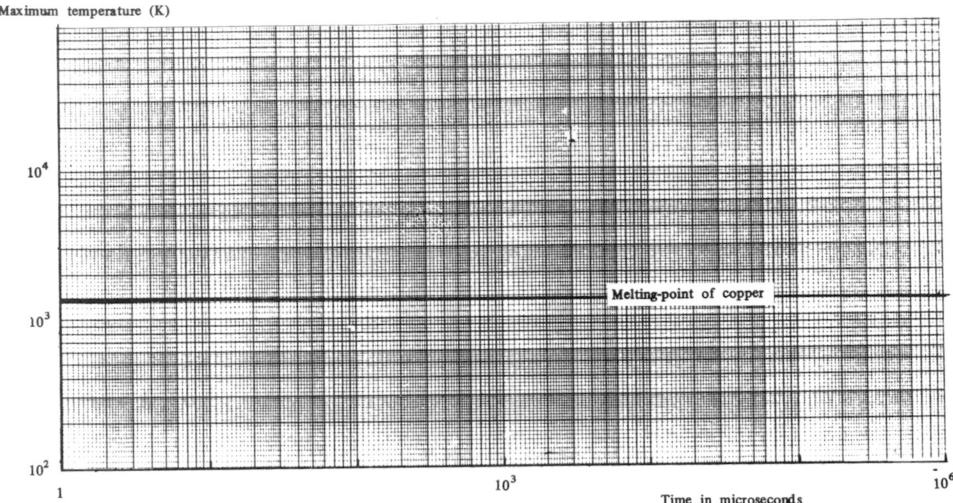

Graph 1.11 Maximum temperature in the conductor against time, various Bi for $\mathcal{S} = 1$.

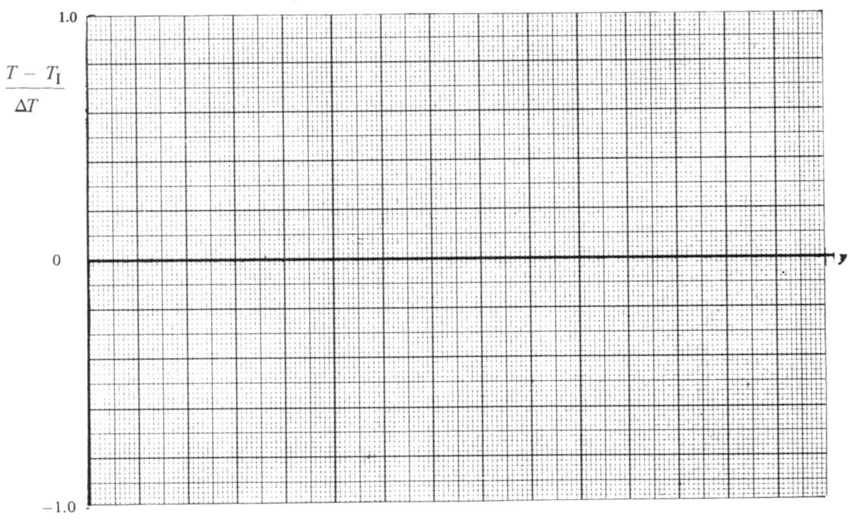

Graph 1.12 Variation of normalised temperature with depth at various numbers of cycles

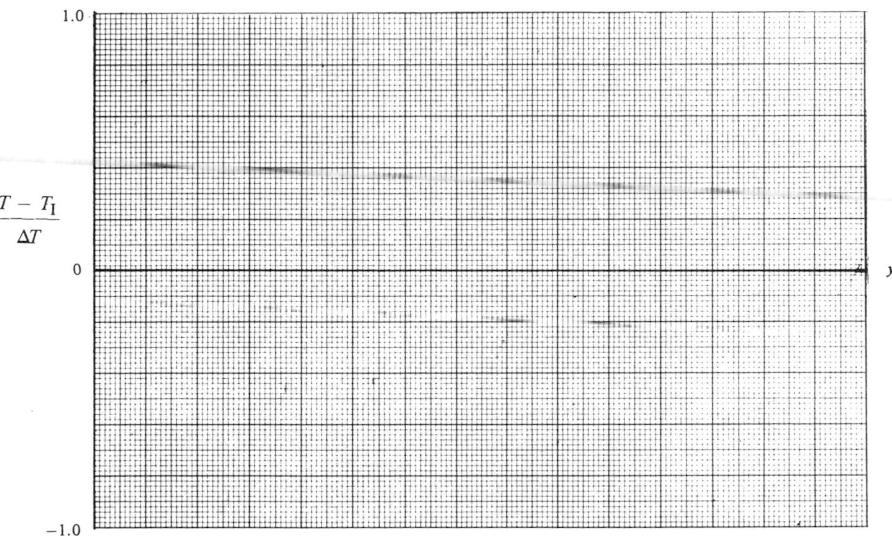

Graph 1.13 Variation of normalised temperature with depth at various steps in a periodic cycle

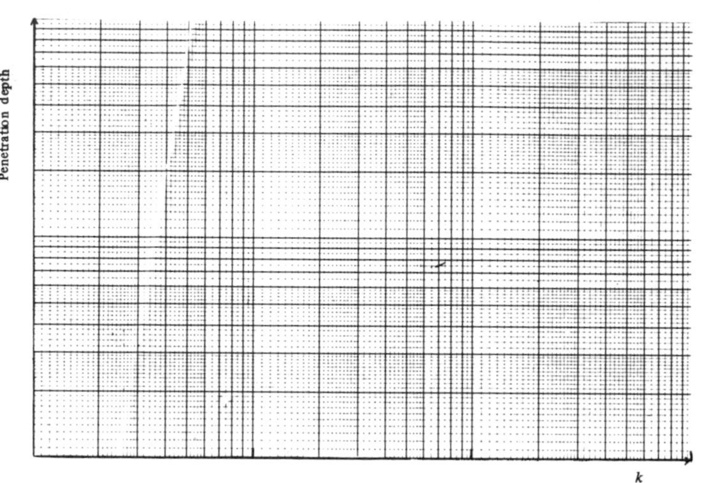

Graph 1.14 Dependence of penetration depth on period

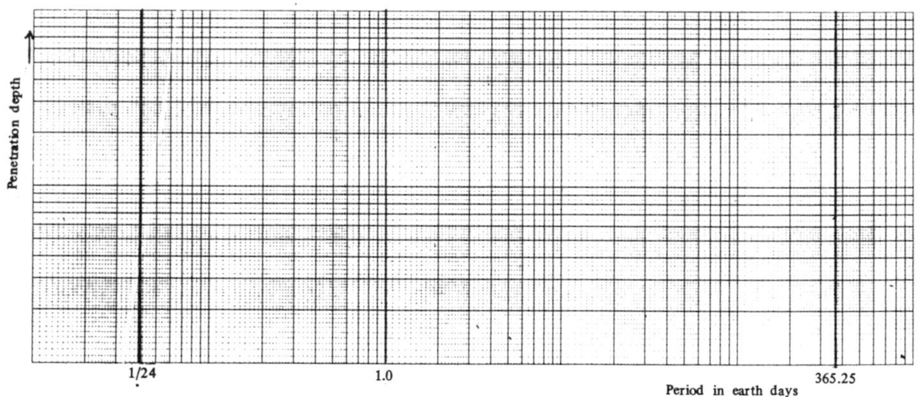

Graph 1.15 Dependence of penetration depth on thermal conductivity.

Applications of TEACH–C
PROBLEM 2
Two-dimensional Conduction Processes

1. Introduction

More often than not, practical heat-conduction situations involve temperature variations and heat flows in more than one direction. Indeed, in general, heat conduction is a three-dimensional phenomenon and it is sometimes necessary to treat it as such.

There exists, however, a class of problems in which temperature variations in one coordinate direction are very much smaller than in the other two, so allowing the problem to be treated as two-dimensional. Examples of such situations are:

- ♦ the extraction of heat from an engine cylinder by annular cooling fins of rectangular section;
- ♦ the generation of heat within a circular-sectioned nuclear fuel rod with an axially-varying fission rate;
- ♦ the quenching of rectangular-sectioned metal ingots whose length is much greater than the maximum cross-sectional dimension;
- ♦ the spread of heat from electrical components mounted on a metal plate with effectively adiabatic surfaces;

and there are many others.

Apart from their practical relevance, such problems are important in developing an understanding of the behaviour of multi-dimensional conduction processes, for it is much easier to represent and interpret temperature variations in a two-dimensional space than in three dimensions. Moreover, from the theoretical point of view this simplification brings no major disadvantages, for it is usually the case that methods developed for two dimensions are capable of straightforward extension to three.

The task of analysis involves solving the governing heat conduction equation, which is typically written in terms of either plane Cartesian coordinates, i.e.

$$\rho c_v \frac{\partial T}{\partial t} - \frac{\partial}{\partial x}\left\{k \frac{\partial T}{\partial x}\right\} - \frac{\partial}{\partial y}\left\{k \frac{\partial T}{\partial y}\right\} - s = 0 \qquad (2.1)$$

or cylindrical-polar ones:

$$\rho c_{\mathrm{v}}\, r \frac{\partial T}{\partial t} - \frac{\partial}{\partial x}\left\{ \kappa r \frac{\partial T}{\partial x} \right\} - \frac{\partial}{\partial y}\left\{ kr \frac{\partial T}{\partial y} \right\} - rs = 0 \tag{2.2}$$

depending on the situation. Analytical solutions of these equations are available (see for example [refs. 2.1 to 2.3]) and it is therefore possible to use them, where they exist, as checks of numerical procedures like that contained in TEACH–C (one such check will be described in LESSON 1 below). In general however the analytical techniques are applicable only in very restricted circumstances, and the solutions which they produce, being often in the form of infinite series, may in any case require a computer to evaluate. It is numerical procedures which are therefore achieving dominance as general-purpose tools.

2. Programming: PROBLEM 2 Base Case

Instructions will be given in this section on how to adapt TEACH–C to the solution of eqns. (2.1) or (2.2) for various circumstances or 'LESSONs' as we shall call them from now on. In preparation for this, we shall establish a basic situation or 'Base Case' on which the individual LESSONs can be regarded as variations. The case is that of a long rectangular bar of width w and height h, initially at a uniform temperature: suddenly a different temperature is imposed along one side, while the other sides are held at the initial value, as illustrated in fig. 2.1 below. It is assumed that the plane of calculation, shown shaded, is sufficiently remote from the ends of the bar that the temperature distribution is independent of the axial coordinate z.

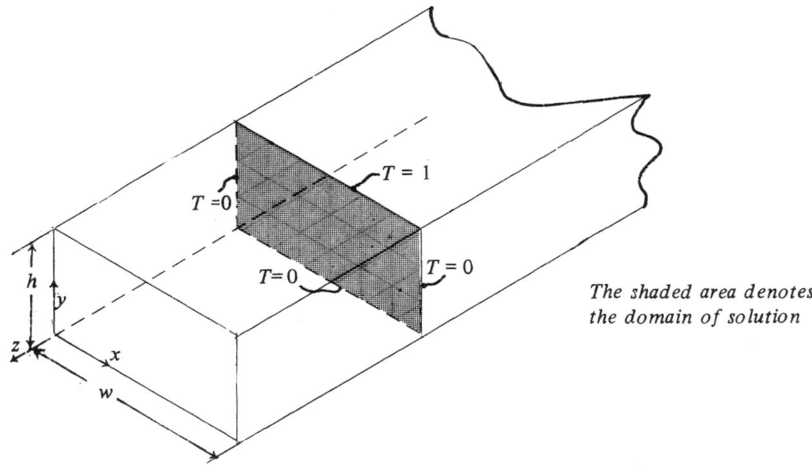

Fig. 2.1 Illustration of PROBLEM 2 Base Case, showing surface temperature distribution for $t > 0$

This situation is identical to that of the illustrative example solved by the standard version of TEACH–C.* Accordingly nothing is required to set up the present Base Case apart from arranging the appropriate output headings. This change is specified in table 2.1 below, which is followed by a listing of the UPDATE which effects it.

TABLE 2.1 Programming instructions for PROBLEM 2 Base Case

Subroutine	Chapter	Changes
CONTRO	4	◆ Provide Format for PROBLEM heading

PROBLEM 2 Base Case UPDATE for effecting the change specified in Table 2.1.

```
*IDENT,PROBLEM2
*D,CONTRO.184,185
 2900 FORMAT(1H1,17X,47HPROBLEM 2  TWO DIMENSIONAL CONDUCTION PROCESSES//
```

3. LESSONs

3.1 LESSON 1 Numerical tests for rectangular bar with prescribed surface temperature.

3.1.1 *Physical background*

In this LESSON we examine a situation similar to that of fig. 2.1, but with the imposed variation of temperature along the upper surface now of the form:

$$T = \sin(\pi x/w) \tag{2.3}$$

This situation might arise if, for example, the bar were part of the insulation matrix of an electrical transformer subjected to a sudden increase in load, in which case the much higher thermal conductivity and heat capacity of the surrounding electrical conductors would cause their temperatures to adjust quickly to the change in current and thereafter remain effectively fixed relative to that of the insulation.

Apart from the practical aspect, this situation will also be used to illustrate two important facets of the numerical solution procedure, namely accuracy

---

* See pages 32–45 above

and economy. The assessment of accuracy is facilitated by the fact that the relevant form of the heat conduction eqn. (2.1) admits, for the temperature distribution of eqn. (2.3), a closed analytical solution in the steady state (which is why this particular distribution has been specified).

It is:

$$T = \frac{\sin(\pi x/w)\,\sinh(\pi y/w)}{\sinh(\pi h/w)} \tag{2.4}$$

obtained by the method of separation of variables. Comparisons will be made between this solution and numerical predictions obtained with various (uniform) grids. No analytical solution is available for comparison in the transient phase of the calculations, but this aspect of the numerical method is examined in PROBLEM 1.

The economy of the procedure refers to the cost of making the calculations, which in turn is a function of the computing time required. The latter is governed by a number of factors, which include:

(*a*) *The fineness of the grid- and time-intervals*:- the maximum allowable values of these are dictated by accuracy requirements, but if they are made smaller than is needed for acceptable accuracy, unnecessary additional costs will be incurred.

(*b*) *The method of approaching the steady state*:- if, as is often the case, only the steady-state solution is of interest, it is possible to bypass the transient phase, and so reduce computing times, by setting the control parameter INTIME = ·FALSE· (this, as described in the TEACH–C manual, has the effect of suppressing the calculation of the $\partial T/\partial t$ term and proceeding to the final solution in a single sequence of iterations).

There are many other factors influencing both economy and accuracy, but here attention will be focused on those just outlined.

3.1.2 *Objectives*

The specific objectives of the LESSON are as follows:

(*a*) To determine, for a square-sectioned bar, the dependence of the accuracy and cost of the TEACH–C predictions on the fineness of the grid, when the program is operated in the steady-state mode.

(*b*) To compare and assess the relative costs of obtaining the steady-state solution by either a full transient calculation, or proceeding directly to the steady state.

3.1.3 *Method*

To achieve the first objective, set TEACH–C to operate in the steady-state mode and make runs for various uniformly-spaced grids, obtained by ascribing NI and NJ equal values according to the following table. Simultaneously alter the values (IMON, JMON) of the temperature-monitoring location to the recommended

TABLE 2.2 Recommended values of program parameters for the grid explorations

NI *and* NJ	IMON *and* JMON
8	4
10	5
12	6
14	7

settings. Determine for each grid: the error, as characterised by the % deviation from the exact solution at the grid location nearest to the centre of the bar; and the 'cost factor', defined as the product of the total number of internal grid nodes [= (NI–2)(NJ–2)] and the total number of iterations required, the whole being multiplied by an arbitrary factor to yield convenient numbers.

For the second objective, make one further run with the program operating in the unsteady mode (i.e. INTIME = ·TRUE·) setting NI = NJ = 12, MAXSTP = 16 and with DT = 300 s initially, and then increased by a factor of 1.5 at each time step. Compare the cost factor with that for the steady-state run for the same grid.

3.1.4 *Programming instructions*

The main changes required to the Base Case are alteration of the temperatures on the top boundary to correspond to eqn. (2.3) and insertion of instructions for calculating the cost factor and evaluating the exact temperature distribution from eqn. (2.4). Table 2.4 summarises all the changes required and Table 2.3 defines the new FORTRAN variables introduced. These are followed by listings of UPDATEs for effecting these changes and then making the necessary RUNs

TABLE 2.3 New FORTRAN variables for LESSON 1

Variable	Quantity	Significance
COST		Computing cost factor
ERR		Percentage error at location (IMON, JMON)
NITOT		Total number of iterations required

TABLE 2.4 Programming instructions for LESSON 1

Subroutine	Chapter	Changes
CONTRO	0	◆ Insert heading for exact solution
	1	◆ Set MAXSTP, SORMAX, DT and NITOT
		◆ Set INTIME to ·FALSE·
	2	◆ Impose upper surface temperature variation
	3	◆ Increase NITOT inside iteration loop
	4	◆ Compute and print cost factor, exact solution and error
		◆ Provide formats for output

PROBLEM 2 LESSON 1 UPDATE for effecting the changes specified in table 2.4

```
*IDENT,P2L1
*D,CONTRO.43
     1     /'        ','EXACT ','SOLUTI','ON     ',2*'       '/
*D,CONTRO.92
      MAXSTP=16
*D,CONTRO.97,98
      SORMAX=0.005
      DT=0.0
*D,CONTRO.100
      INTIME=.FALSE.
*D,CONTRO.116
      T(I,NJ)=SIN(PI*X(I)/W)
*I,CONTRO.133
      CALL SECOND(T1)
      WRITE(6,3901)T1
*I,CONTRO.137
      IF (NSTEP.NE.1) DT=DT*1.4
*I,CONTRO.181
      CALL SECOND(T2)
      DELTAT=T2-T1
      COST=DELTAT*35.
      WRITE(6,4901)T2,DELTAT,COST
C-----CALCULATION OF EXACT STEADY-STATE SOLUTION AND ERROR
      HW=PI*H/W
      SINHH=0.5*(EXP(HW)-EXP(-HW))
      TIJ=T(IMON,JMON)
      DO 4101 I=2,NIM1
      XW=PI*X(I)/W
      DO 4101 J=2,NJM1
      YW=PI*Y(J)/W
      SINHY=0.5*(EXP(YW)-EXP(-YW))
 4101 T(I,J)=SIN(XW)*SINHY/SINHH
      CALL PRINT(1,1,NI,NJ,IT,JT,X,Y,T,HEDS)
      ERR=ABS((T(IMON,JMON)-TIJ)/T(IMON,JMON))*100.
      WRITE (6,4902) IMON,JMON,ERR
*I,PROBLEM2.1
     1   /17X,47HLESSON 1    NUMERICAL TESTS FOR RECTANGULAR BAR
     1   /29X,35HWITH PRESCRIBED SURFACE TEMPERATURE//
*I,CONTRO.201
 3901 FORMAT(///5X,16H STARTING TIME= ,F8.4,6H SECS.//)
 4901 FORMAT(//5X,17H FINISHING TIME= ,F8.4,6H SECS.//,5X,25H TOTAL CALC
     1ULATION TIME= ,F8.4,6H SECS.//5X,16H COMPUTING COST=,F8.2,6H UNITS
     1//)
 4902 FORMAT (//5X,31H PERCENTAGE ERROR AT LOCATION (,I2,1H,,I2,2H)=,
     1 F8.4)
```

AN UPDATE of the following form will be required to perform the RUNs of LESSON 1

```
*IDENT,P2L1R1
*D,CONTRO.57,58
         NI=10
         NJ=10
*D,CONTRO.84,85
         IMON=5
         JMON=5
*D,P2L1.4
         DT=300.0
*D,P2L1.6
         INTIME=.TRUE.
```

3.1.5 *Analysis of results*

(*i*) Insert on graph* 2.1 the variation of the % error at the centre of the bar versus NI: then determine the grid which is required to yield an error of about 0.1%.

(*ii*) Plot on the same graph the cost of each RUN, in terms of the cost factor. Comment on the advisability of employing grids which are finer than is required for the accuracy specified above.

(*iii*) Determine, from the computing costs recorded for part (*b*), the relative costs of obtaining a steady-state solution for this particular conduction problem by the two alternative routes. What general conclusions may be drawn from these?

3.2 LESSON 2 Investigation of the superposition principle

3.2.1 *Physical background*

The term 'superposition' refers to a technique which exploits the linearity of the heat-conduction equation prevailing in certain circumstances to generate solutions of complex problems by linearly combining those of simpler ones. The circumstances in question are those in which the material properties are temperature-independent and heat sources (if present) and boundary conditions are at most linearly dependent on temperature.

The superposition principle has seen widespread application in the solution of conduction problems by analytical techniques: see for example refs. (1.1)–(1.3).

The purpose of this LESSON is to investigate the validity and conditions of applicability of superposition by use of numerical solution techniques. This exercise will be carried out for a material with uniform and constant

* Graphs will be found on pages 127–135

thermal conductivity k_0 and for one whose conductivity varies with temperature according to:

$$k = k_0 \ (1 + 5T^2) \tag{2.5}$$

The approach will be to solve for the temperature distributions in the three square-sectioned bars, **A**, **B** and **C** of fig. 2.2, which differ only in respect of the boundary temperatures, and to then compare the solutions in the manner indicated below. In each case the calculation will be a transient one, with the interior temperatures being zero initially.

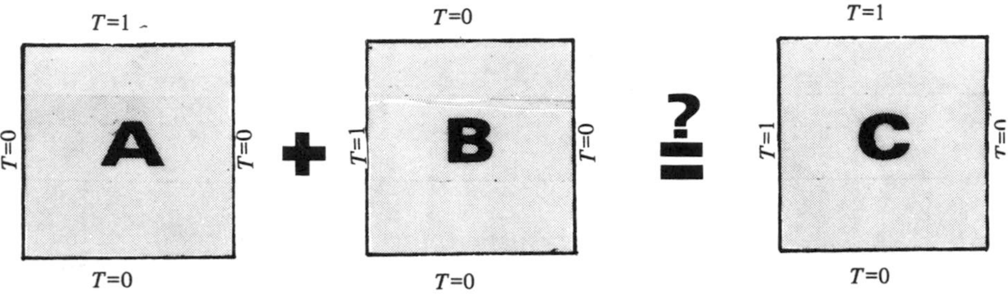

Fig. 2.2 **Illustration of the situation examined in the superposition investigation**

3.2.2 *Objectives*

The objectives of the LESSON will be to assess the validity of the superposition principle by determining whether the equality of fig. 2.2 holds for:

(*a*) constant thermal conductivity;

(*b*) temperature-dependent conductivity according to eqn. (2.5).

3.2.3 *Method*

The material properties and dimensions of the bars are of no importance in these calculations, provided that they are the same for each bar: thus they have been assigned arbitrary values.

For objective (*a*) obtain solutions for each of the three situations by assigning appropriate values to the boundary temperatures and then operating TEACH–C in the unsteady-state mode, with NI = NJ = 12 and DT and MAXSTP set at 0.05s and 15 respectively.

Objective (*b*) requires similar RUNs, but with the temperature-dependent thermal conductivity inserted.

3.2.4 *Programming instructions*

The situations examined in this LESSON differ from that of the PROBLEM 2 Base Case only in respect of the values assigned to the boundary temperatures, the insertion of the temperature-dependent thermal conductivity and other minor details. The conductivity formula (2.5) should be inserted in subroutine PROPS and the specification INPRO = ·TRUE· should be made, so as to cause PROPS to be called at each iteration. Table 2.5 summarises all the necessary changes and this is followed by listings of the UPDATEs for setting up the LESSON and making the RUNs.

TABLE 2.5 Programming instructions for LESSON 2

Subroutine	Chapter	Changes
CONTRO	1	♦ Specify (arbitrary) material properties
		♦ Set MAXSTP, SORMAX and DT.
	2	♦ Assign boundary temperatures
	4	♦ Modify formats for LESSON heading
PROPS	0	♦ Insert temperature COMMON block

PROBLEM 2 LESSON 2 UPDATE for effecting the changes specified in Table 2.5.

```
*IDENT,P2L2
*D,CONTRO.87,89
      TCON=1.0
      CV=1.0
      DENSIT=1.0
*D,CONTRO.92
      MAXSTP=15
*D,CONTRO.97,98
      SORMAX=0.005
      DT=0.05
*D,CONTRO.111
      TTOP=1.0
*I,PROBLEM2.1
     1/16X,54H LESSON 2       INVESTIGATION OF SUPERPOSITION PRINCIPLE //
*I,PROPS.12
     1/TEMP/T(22,22),TOLD(22,22)
```

An UPDATE of the following form will be required to perform the RUNs of LESSON 2:

```
*IDENT,P2L2R1
*D,CONTRO.1Ø3
      INPRO=.TRUE.
*D,P2L2.7
      TTOP=Ø.Ø
*D,CONTRO.113
      TLEFT=1.Ø
*D,PROPS.18
      GAMH(I,J)=TCON*(1.Ø+5.Ø*T(I,J)**2)
```

3.2.5 *Analysis of results*

(*i*) For the case of constant conductivity insert in graph * 2.2 the variation with time of the temperature at the monitoring location for situations **A**, **B** and **C**, labelling each curve appropriately. Plot also $T_A + T_B$ and comment on the results in the light of the superposition principle. Make spot checks to determine whether the same behaviour prevails at other points in the grid.

(*ii*) Perform the same analysis for the variable-conductivity case, using graph 2.3 for the plotting.

(*iii*) Consider the heat flows through the surfaces of the bars and deduce whether they too should obey the superposition principle, and if so, under what circumstances.

3.3 LESSON 3 Analysis of fin heat transfer

3.3.1 *Physical background*

Heat transfer can be cheaply, and often very satisfactorily, augmented by simply using fins to extend the available surface area. There are thus many engineering applications: for example fins are employed on all air-cooled reciprocating engines; they frequently appear on all forms of electrical machinery and components of electronic circuits; and they are an essential feature of the modern compact heat exchanger.

In the present LESSON we examine the heat-transfer performance of fins of rectangular cross section, with thickness h and length w, as depicted in fig. 2.3. Two shapes of fin are shown in the figure, namely plane and radial, but we shall be concentrating in this LESSON on the former ** which may be regarded as a special case of the latter, obtained by causing the inner radius r_0 to be large. The relevant form of the heat conduction eqn. (2.1) to be solved is, for the case of uniform thermal conductivity:

$$\frac{\partial^2 T}{\partial x^2} + \frac{\partial^2 T}{\partial y^2} = 0. \qquad (2.6)$$

We shall presume that the temperature T_B of the base of the fin is known. This is often so in practice – for example, in the case of the air-cooled engine block temperature. We shall also suppose that the heat transfer coefficient α between the fin surface and the surroundings of temperature T_F is also known and uniform. In practice, it will not be known (nor necessarily uniform) and will require to be determined from convective flow considerations. However, it will suffice for the present purposes to simply take α as specified.

* Graphs will be found on pages 127–135

** Because closed analytical solutions exist for limiting cases, with which we can compare.

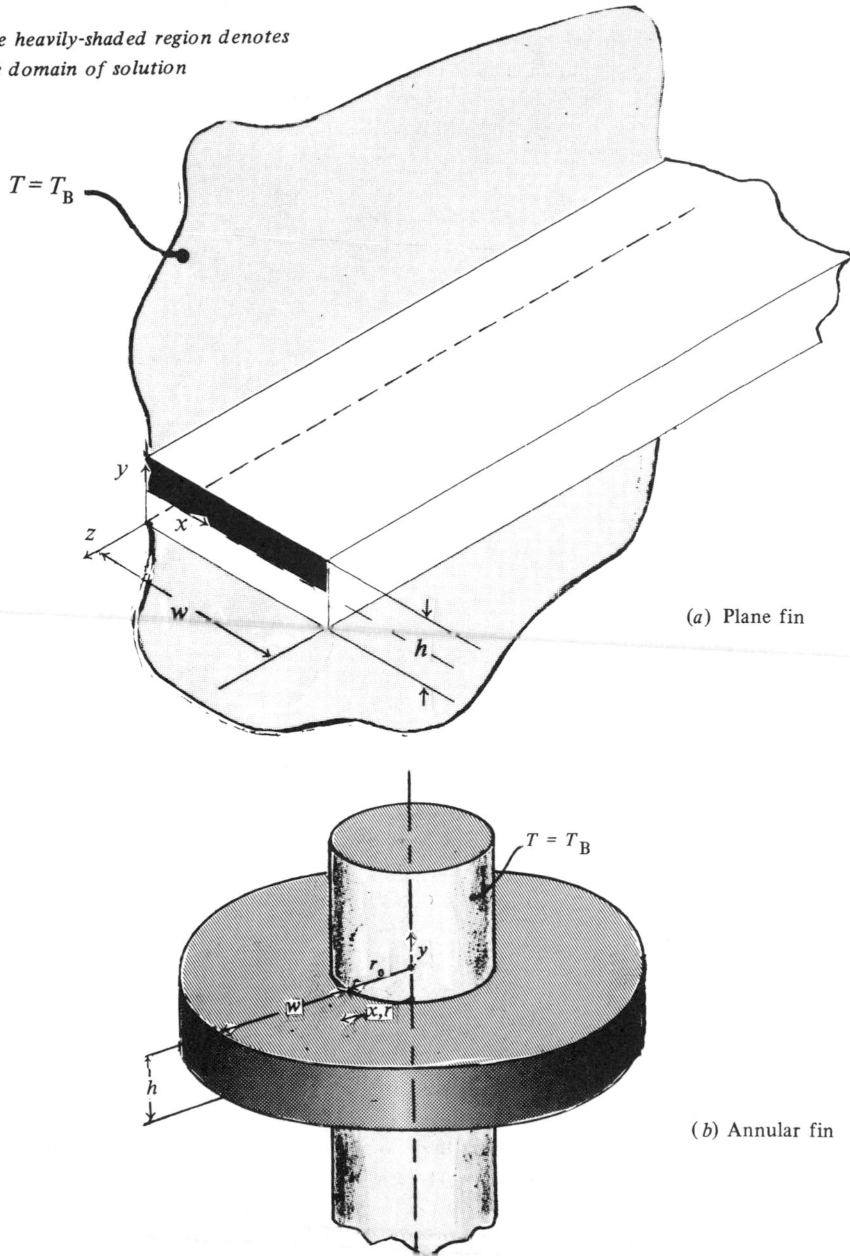

Fig. 2.3 Illustration of two types of fin having rectangular cross-section

In the foregoing circumstances inspection of the governing differential equation and boundary conditions reveals that two dimensionless parameters govern the fin heat transfer, namely the aspect ratio $A = h/w$ and a Biot number $\mathrm{Bi} = \alpha L/k$, where the characteristic length L may be either h or w. Since the Biot number measures the ratio of the resistance to conduction within the fin to the resistance to convection from its surface, and since we shall be concerned with determining when the cross-fin conduction resistance causes significant temperature gradients in that direction, we shall use the definition $\mathrm{Bi} = \alpha h/k$.

One-dimensional fin analysis (the 'thin fin' approximation) is widely used by engineers to estimate fin heat transfer because of the relative ease of analytical solution. Schneider [ref. (2.3)] gives the solution for the heat transfer rate $\dot{q}_{1D}$ from the plane fin which is

$$\frac{\dot{q}_{1D}}{\alpha(T_B - T_F)} = \sqrt{2\mathrm{Bi}}\;\frac{\sqrt{\mathrm{Bi}/2} + \tanh\left(\sqrt{2\mathrm{Bi}}\, w/h\right)}{1 + \sqrt{\mathrm{Bi}/2}\,\tanh\left(\sqrt{2\mathrm{Bi}}\, w/h\right)} \tag{2.7}$$

The fin efficiency ϵ is often used to measure the heat-transfer performance. It is defined by:

$$\epsilon = \frac{\text{'actual' heat transfer from fin, } \dot{q}}{\text{heat transfer from an idealised fin of temperature } T_B \text{ throughout, } \dot{q}_{IDEAL}}$$

In the present LESSON we shall obtain $\dot{q}$ and ϵ from the numerical solution of eqn. (2.6) and compare the former with $\dot{q}_{1D}$ to determine the conditions of validity of eqn. (2.7).

3.3.2 *Objectives*

The objectives of the LESSON are thus:

(*a*) To calculate the steady-state heat transfer from, and temperature distribution in, a cooling fin for various values of the governing parameters.

(*b*) To identify conditions for which the fin performance is essentially that given by a one-dimensional analysis.

(*c*) To examine the dependence of the fin efficiency on the governing parameters.

3.3.3 *Method*

Perform calculations for aspect ratios of 5, 10, and 20 in each case for Biot numbers of 10^{-2}, 10^{-1}, 1 and 10. Determine the fin efficiency ϵ, and the ratio of the computed to 'one-dimensional' heat transfers for each RUN.

The values of the quantities w, h, α, k, T_B and T_F which you use are arbitrary. However, from the point of view of introducing a measure of physical reality you might choose to make $h = 2 \times 10^{-3}$ m say, and $k = 52$ J/msK, the value for steel, and let $T_B = 100°$C and $T_F = 0°$C. The aspect ratio and Biot number may then be specified through w and α respectively.

The recommended grid for these calculations is uniformly distributed in the y (i.e. cross-fin) direction with NJ = 7 and expands at a constant rate* in the x direction, with NI = 12 and the expansion factor FEXPX varied according to the value of Bi (but *not* A) as follows:

TABLE 2.6 Recommended values of the grid expansion factor FEXPX for different Biot numbers

Bi	FEXPX
10	1.8
1	1.8
10^{-1}	1.4
10^{-2}	1.2

This type of grid is employed because in general the temperature gradient $\partial T/\partial x$ along the fin decreases with x and increases with Bi.

3.3.4 *Programming instructions*

The further changes required to produce the LESSON 3 UPDATE are concerned with inserting the boundary conditions and arranging for appropriate output. Table 2.7 lists the new FORTRAN variables, table 2.8 summarises all the changes required, and listings are then provided of the UPDATEs themselves.

TABLE 2.7 New FORTRAN variables for LESSON 3

Variable	Quantity	Significance
ALPHA	α	Heat transfer coefficient
BIOT	Bi	Biot number
QEF	ϵ	Fin efficiency
QEFF		Ratio of numerically-predicted heat transfer to 1-dimensional heat transfer rates
QTOT	$\dot{q}$	Numerically-predicted heat transfer rate
Q1D	$\dot{q}_{1D}$	1-dimensional heat transfer rate
TFIN	T_B	Temperature at base of fin
TFLUID	T_F	Fluid temperature

* According to a built-in formula provided in the basic TEACH–C program, listed in the lines CONTRO.63 to CONTRO.78 of the code (see page 57).

TABLE 2.8 Programming instructions for LESSON 3

Subroutine	Chapter	Changes
CONTRO	0	◆ Insert ALPHA, TFIN and TFLUID in common block.
	1	◆ Assign values to NI, NJ, H and W.
		◆ Set ALPHA, TFIN and TFLUID
		◆ Calculate Biot number
		◆ Set MAXIT, NITPRI, SORMAX and DT.
		◆ Specify steady-state calculation by setting INTIME to ·FALSE·
	2	◆ Initialise interior temperature field.
	4	◆ Calculate heat transfer from fin and efficiency
		◆ Calculate heat transfer from 1-dimensional eqn. (2.7)
		◆ Provide formats for LESSON output
PROMOD	0	◆ Insert common block as in CONTRO
	1	◆ Make south boundary a plane of symmetry
		◆ Apply convective boundary conditions at north and east boundaries

PROBLEM 2 LESSON 3 UPDATE for effecting the changes specified in table 2.8

```
*IDENT,P2L3
*I,CONTRO.37
     1/P2L3/ ALPHA,TFIN,TFLUID
*D,CONTRO.57,60
      NI=12
      NJ=7
      W=0.01
      H=0.002
      H=H/2.
*D,CONTRO.85
      JMON=4
*I,CONTRO.89
C-----ENTER FIN TEMP., FLUID TEMP., AND ALPHA.
      ALPHA=2.6E4
      TFIN=100.0
      TFLUID=0.0
      BIOT=2.*ALPHA*H/TCON
*D,CONTRO.91
      MAXIT=300
*D,CONTRO.93
      NITPRI=50
*D,CONTRO.97,98
```

```
C-----DECREASE SORMAX TO INCREASE ACCURACY FOR LOW BIOT VALUES.
      SORMAX=1.E-6
      DT=0.0
*D,CONTRO.100
      INTIME=.FALSE.
*D,CONTRO.111,120
      RATT1=TCON/(ALPHA*(YV(NJ)-Y(NJM1)))
      RATT2=TCON/(ALPHA*(XU(NI)-X(NIM1)))
      DO 2001 I=1,NI
      DO 2001 J=1,NJ
 2001 T(I,J)=TFIN
*D,CONTRO.124,127
      SNORM=ABS(TCON*(TFIN-TFLUID))*AMAX1(H/W,W/H)
*I,CONTRO.128
      H=2.*H
*D,CONTRO.129
      WRITE(6,2900)H,W,CV,TCON,DENSIT,ALPHA,BIOT,TFIN,TFLUID,SNORM,NI,NJ
      H=0.5*H
*D,CONTRO.149
      DO 3201 I=2,NIM1
      T(I,NJ)=(RATT1*T(I,NJM1)+TFLUID)/(RATT1+1.)
 3201 T(I,1)=T(I,2)
      DO 3202 J=1,NJ
 3202 T(NI,J)=(RATT2*T(NIM1,J)+TFLUID)/(RATT2+1.)
*I,CONTRO.181
C-----TO DETERMINE THE 1-D SOLUTION
      BISR=SQRT(BIOT)
      H=2.*H
      HTAN=SINH(BISR*W/H*SQRT(2.))/COSH(BISR*W/H*SQRT(2.))
      QST=SQRT(2.)*BISR*((SQRT(0.5)*BISR+HTAN)/(1.+
     1SQRT(0.5)*BISR*HTAN))
      Q1D=(TFIN-TFLUID)*TCON*QST
C-----CALCULATE ACTUAL HEAT TRANSFER.
      QTOT=0.0
      NJM1=NJ-1
      DO 4101 J=2,NJM1
      Q=2.*(TFIN-T(2,J))*SNS(J)/(X(2)-XU(2))
 4101 QTOT=QTOT+Q
      QTOT=QTOT*TCON
      QEFF=QTOT/Q1D
C-----CALCULATE HEAT TRANSFER WITH FIN AT TFIN, I.E. PERFECT.
      QPER=ALPHA*(TFIN-TFLUID)*(2.*W+H)
      QEF=QTOT/QPER
      WRITE (6,4901) QEF,Q1D,QTOT,QEFF
*I,PROBLEM2.1
     1/16X,49H LESSON 3      CONDUCTION IN A FIN WITH PRESCRIBED
     1/30X,25HHEAT TRANSFER COEFFICIENT//
*D,CONTRO.191
     1/16X,40H HEAT TRANSFER COEFFICIENT, ALPHA -----=,G10.3
     1/16X,40H BIOT NUMBER, BIOT --------------------=,G10.3
     1/16X,40H INITIAL FIN TEMPERATURE, TFIN --------=,G10.3
     1/16X,40H FLUID TEMPERATURE, TFLUID ------------=,G10.3
*I,CONTRO.201
 4901 FORMAT (//28H FIN EFFECTIVENESS --------=,1PE12.3
     1/28H ONE DIMENSIONAL HEAT FLUX =,E12.3
     1/28H REAL HEAT FLUX -----------=,E12.3
     1/28H RATIO OF FLUXES, REAL/1D -=,E12.3)
*I,PROMOD.15
     1/P2L3/ ALPHA,TFIN,TFLUID
*D,PROMOD.26,28
C-----INSERT CONVECTIVE NORTHERN BOUNDARY
      RN=1./(ALPHA*SEW(IL))+1./DN
      SU(NJM1)=SU(NJM1)+TFLUID/RN
      SP(NJM1)=SP(NJM1)-1./RN
*D,PROMOD.30,32
C-----SYMMETRIC SOUTH BOUNDARY
*D,PROMOD.39,40
C-----INSERT CONVECTIVE EASTERN BOUNDARY.
      RE=1./(ALPHA*SNS(J))+1./DE
      SU(J)=SU(J)+TFLUID/RE
      SP(J)=SP(J)-1./RE
```

An UPDATE of the following form will be required to perform the RUNs of LESSON 3

```
*IDENT, P2L3R1
*D, P2L3.4
        W=Ø.Ø1
*D, CONTRO.61
        FEXPX=1.8
*D, P2L3.9
        ALPHA=2.6E5
```

3.3.5 *Analysis of results*

(*i*) For each of the three aspect ratios, plot on the separate graphs * 2.4(*a*), (*b*) and (*c*) the temperatures along the $y = h$ symmetry plane with the Biot number as a parameter. Are the effects of aspect ratio and Biot number in accordance with your expectations? Explain what information emerges from the plot which would be of assistance in designing an economical fin (i.e. one formed from a minimum mass of material)?

(*ii*) Insert in graph 2.5 the temperature distribution across the $A = 10$ fin at $x/w \approx 0.1$, for each Biot number. Discuss the relationship between the Biot number and the degree of departure of the temperature profiles from one-dimensional behaviour.

(*iii*) On graph 2.6 plot the fin effectiveness against Biot number for the three aspect ratios. Explain how the behaviour of the effectiveness accords the implications of graphs 2.4. Discuss the philosophy of preparing plots like that of graph 2.6 with the aid of the computer within the context of their benefit to practising engineers.

(*iv*) Insert in graph 2.7 the variation of the ratio of the numerically-predicted heat transfer to the one-dimensional analytical value given by eqn. (2.7) with Biot number, for the three aspect ratios. Explain the results and deduce the range of Biot number for which the one-dimensional analysis is satisfactory. What is the dependence of the heat transfer ratio on aspect ratio? Does this accord with your understanding of the significance of Bi, as defined in this LESSON?

3.4 LESSON 4 Temperature distribution in a nuclear fuel rod

3.4.1 *Physical background*

In this LESSON we consider steady-state heat conduction in a circular-sectioned rod of diameter $2h$ and length $2w$ in which heat generation is present. Specifically, the LESSON examines the kind of heating that may typically arise in a pressurised steam-generating nuclear power reactor. The rod would be a cylinder of uranium dioxide, illustrated in fig. 2.4, which

* Graphs will be found on pages 127–135

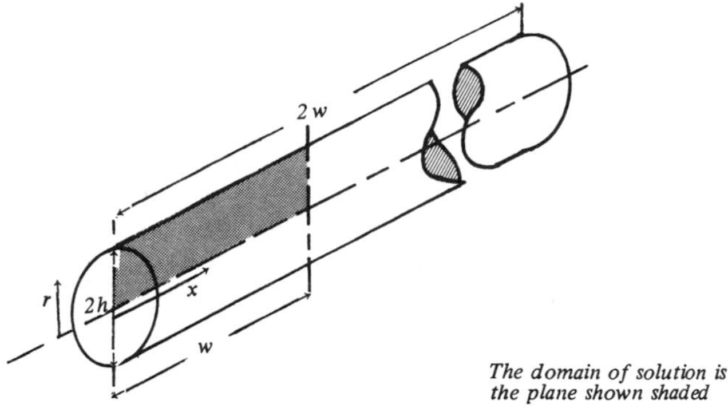

Fig. 2.4 Illustration of nuclear fuel rod, showing solution domain

serves as a fuel by undergoing controlled fission. The fuel would be encased in a zircalloy container whose outer surface would be cooled by a flow of boiling water. In this physical situation the rate of heating along the rod is not uniform but increases with the density of the neutron flux, which is usually taken to vary sinusoidally along the fuel pin. Consequently we assume the local volumetric heat generation rate s to be given by:

$$s = \dot{q}'''_{max} \sin(\pi x/2w) \tag{2.8}$$

where $\dot{q}'''_{max}$ is proportional to the fission rate.

It is a reasonable approximation to treat the surface temperature of the nuclear fuel as constant; a value of 600°C is adopted, which is also the temperature assigned to the end faces. An important question then arises: what is the maximum allowable heat release such that the maximum temperature within the uranium dioxide should not exceed 900°C (which is close to the safe maximum temperature for this fuel)?

The present LESSON is concerned with answering this question and others relating to the temperature distribution within the rod.

3.4.2 *Objectives*

The objectives of the LESSON are:

(*a*) To find the maximum allowable heat release from the fuel rod so that the peak temperature within the uranium dioxide should not exceed 900°C.

(*b*) To determine the relative importance of axial and radial conduction (and hence the validity of analyses based on a one-dimensional conduction assumption) in the rod as prescribed in (*a*).

(*c*) To explore the effect of the length : diameter ratio of the rod on the interior temperature distribution.

3.4.3 *Method*

The bulk of the calculations are made for a rod 0.02m in diameter and 2m long, having the material properties of uranium dioxide (we ignore, for convenience, the presence of the zircalloy sheath, whose thermal resistance will in any case be small). Because the heating rate given by eqn. (2.8) is symmetrical about $x = w$, we consider only the region $0 \leqslant x \leqslant w$ shown in fig. 2.4 and we superimpose a uniform grid with NI = 12 and NJ = 8. The residual-source normalisation factor will be taken as $\dot{q}'''_{max} h^2 w/2$ which is (approximately) the total heat generation per unit angle.

For the explorations required for objective (*a*), RUNs should be made with successively larger rates of heat generation, obtained by varying $\dot{q}'''_{max}$ until the maximum allowable temperature has been exceeded; then find the limiting heat generation rate by interpolation. Use as a starting rate the value given in the UPDATE listing, and then increase it successively by factors of 2.

For (*b*) make two further RUNs of the program, using the highest heat rate examined in (*a*). One run should be made with axial conduction suppressed by setting AE(J) and AW(J) to zero on all grid lines and in the other RUN suppress radial conduction by setting the AN(J) and AS(J) to zero.

Finally, the explorations for objective (*c*) should be made for length: diameter ratios of 20:1, 2:1 and 1:1 by holding w constant and varying h. Also, in order to facilitate interpretation of these results a uniform rather than sinusoidally-varying heat generation rate should be prescribed, and the rate should be made inversely proportional to h^2 so as to maintain the temperature level virtually constant.

3.4.4 *Programming*

The main programming tasks are the insertion of the heat source, the specification of the material properties and control parameters, and arranging for suitable printout. These and other changes are summarised in table 2.10, the new FORTRAN variables mentioned therein being defined in table 2.9. There follow the LESSON and RUN UPDATE listings.

TABLE 2.9 New FORTRAN variables for LESSON 4

Variable	Quantity	Significance
S	s	Local volumetric heat generation rate
TI		Initial value of temperature over whole field

TABLE 2.10 Programming instructions for LESSON 4

Subroutine	Chapter	Changes
CONTRO	0	◆ Insert S and W in COMMON block
		◆ Set heading for normalised temperatures
	1	◆ Set INCYLY to .TRUE.
		◆ Assign values to NI, NJ, H, W and S
		◆ Supply material properties of uranium dioxide
		◆ Set DT=0 and INTIME = .FALSE.
	2	◆ Initialise temperature field
		◆ Calculate source-normalisation factor SNORM
	3	◆ Set temperatures along symmetry axis
	4	◆ Calculate and print out normalised temperatures
		◆ Provide formats for output
PROMOD	0	◆ COMMON block as in CONTRO
	1	◆ Make South and West boundaries planes of symmetry
		◆ Insert heat source
PRINT	0	◆ Change heading for radial geometry

PROBLEM 2 LESSON 4 UPDATE for effecting the changes specified in table 2.10

```
*IDENT,P2L4
*I,CONTRO.37
     1/P2L4/ S,W
*D,CONTRO.43
     1      /'        ','NORMAL','IZED T','EMPERA','TURES ','      '/
*D,CONTRO.56,60
      INCYLY=.TRUE.
      NI=12
      NJ=8
      W=2.0
      H=0.1
      S=5.0E07
      S=S*1.0E-4/H/H
*D,CONTRO.84,89
      IMON=6
      JMON=4
C-----MATERIAL PROPERTIES OF URANIUM DIOXIDE
      TCON=9.174
      DENSIT=1.096E04
      CV=3.342E02
*D,CONTRO.98
      DT=0.0
*D,CONTRO.100
      INTIME=.FALSE.
*D,CONTRO.111,120
      TI=600.
      DO 2001 I=1,NI
      DO 2001 J=1,NJ
 2001 T(I,J)=TI
```

```
*D,CONTRO.124,127
      SNORM=S*H*W/2.0
*D,CONTRO.149
C-----UPDATE TEMPERATURES ALONG SYMMETRY AXES
      DO 3201 I=1,NI
 3201 T(I,1)=T(I,2)
      DO 3202 J=1,NJ
 3202 T(NI,J)=T(NIM1,J)
*I,CONTRO.181
C-----CALCULATION OF NORMALIZED TEMPERATURES.
      TNI1=T(NI,1)
      DO 4101 I=1,NI
      DO 4101 J=1,NJ
 4101 T(I,J)=(T(I,J)-TI)/(TNI1-TI)
      CALL PRINT(1,1,NI,NJ,IT,JT,X,Y,T,HEDS)
      SN=S/TCON/(T(NI,1)-TI)
      WRITE(6,4901) SN
*D,CONTRO.186
     1/16X,54H LESSON 4      AXISYMMETRIC CONDUCTION WITH HEAT SOURCE//
     2/16X,40H RADIUS, H ----------------------------=,1PG10.3,2H M
*I,CONTRO.201
 4901 FORMAT(/34X,18HNORMALIZED SOURCE=,1PE12.3)
*I,PROMOD.15
     1/P2L4/ S,W
*D,PROMOD.30,32
*D,PROMOD.38,40
*D,PROMOD.50
C-----HEAT SOURCE DUE TO SINE HEAT FLUX
 1400 PI=4.*ATAN(1.)
      DO 1401 J=2,NJM1
      DV=RY(J)*SEW(IL)*SNS(J)
      SU(J)=SU(J)+S*DV*SIN(PI/2.*X(IL)/W)
 1401 CONTINUE
      RETURN
*D,PRINT.18
      DATA HI,HY/'    I =','R =    '/
```

An UPDATE of the following form will be required to perform the RUNs of LESSON 4.

```
*IDENT,P2L4R1
*D,P2L4.7
      H=Ø.1
*D,P2L4.42
      SU(J)=SU(J)+S*DV
*I,P2L4.43
      DO 14Ø2 J=1,NJ
      AN(J)=Ø.Ø
      AS(J)=Ø.Ø
 14Ø2 CONTINUE
```

3.4.5 *Analysis of results*

(*i*) Plot on graph * 2.8 the variation of maximum temperature with heating rate, characterised by s and hence deduce the limiting rate. Where does the maximum temperature occur? How does it vary with heating rate? Explain.

(*ii*) Insert in graphs 2.9(*a*) and (*b*) the temperature distributions along the I = 12 and J = 1 grid lines respectively for the runs of objective (*b*). Hence deduce the relative importance of axial and radial conduction for these circumstances. Would a one-dimensional analysis provide reasonable accuracy at all locations in the rod?

*Graphs will be found on pages 127–135

(*iii*) Sketch first on the computer printout and then on graphs 2.10(*a*)–(*c*) the isotherms corresponding to values of the normalised temperatures of 0.1, 0.5 and 0.8 for the different length: diameter ratios studied. Summarise and explain the main changes in the shapes of these contours as w/h is altered.

3.5 LESSON 5 The quenching of a metal ingot

3.5.1 *Physical background*

In this LESSON we are concerned with determining the transient evolution of temperature within a long rectangular steel ingot of the form illustrated in fig. 2.1, which is initially at temperature T_I and is suddenly immersed, for heat treatment purposes, in a stirred bath of constant temperature T_F. After immersion of the ingot the boundary condition at the surfaces of the ingot is:

$$\dot{q}''_B = k \left. \frac{\partial T}{\partial n} \right|_B = -\alpha (T_B - T_F) \qquad (2.9)$$

where B refers to the boundary, $\dot{q}''_B$ is the heat flux along the inward-directed normal $\hat{n}$ and α is the heat transfer coefficient. In practice the α may vary with position and time: however for the present purposes, α will be taken as uniform and constant.

It is possible to show that, for these circumstances the non-dimensional temperature $T^* = (T - T_I)/(T_F - T_I)$ after immersion will be a function of:

- the dimensionless independent variables $x_* \equiv x/w$, $y_* \equiv y/h$ and $\text{Fo} \equiv kt/\rho c_v h^2$ (the *Fourier number*, a dimensionless time);
- the parameters $A \equiv w/h$ (aspect ratio) and $\text{Bi} \equiv \alpha h/k$ (*Biot number*)

Here, it should be noted, w and h are respectively defined as the half-width and half-height of the ingot, for later convenience.

The Biot number provides a measure of the ratio of the resistance to heat conduction within the ingot to the resistance to heat convection from its surfaces (there are in fact two Biot numbers, the other being $\alpha w/k$, but this, being equal to the product of Bi and A, is not an independent parameter) and therefore exerts a decisive influence on the quenching process.

Very large values of Bi, achieved by intense stirring, imply that convection offers negligible resistance compared with conduction: thus this situation will be effectively indistinguishable from that where the surface temperature remains constant, as in the situation of fig. 2.1, although here it takes on the uniform value T_F of the fluid. The other extreme is when Bi becomes very small; then the convection resistance dominates, and the temperature everywhere within the ingot is given by:

$$T^* = \exp(-\text{Bi} \cdot \text{Fo}) \qquad (2.10)$$

a result given in standard textbooks on heat conduction.

Limiting cases also exist for extreme values of the aspect ratio A. Thus, for example, when A becomes very large, the temperature will become effectively independent of $x*$, except in a very narrow layer adjacent to the short sides.

The purpose of this LESSON is to determine the temperature behaviour for intermediate as well as extreme values of Bi and A.

3.5.2 *Objectives*

The specific objectives are thus to determine the dependence of the temporal and spatial variations of temperature within the ingot on:

(*a*) The Biot number, at fixed aspect ratio

(*b*) The aspect ratio, at fixed Biot number.

3.5.3 *Method*

The calculations will be performed for a steel ingot of height $2h$ = 0.2m. The symmetry of the problem about the planes $x = w$ and $y = h$ (fig. 2.5) allows the solution domain to be confined to just one quadrant, shown shaded in the diagram. A uniform grid will be employed with NI = NJ = 8;

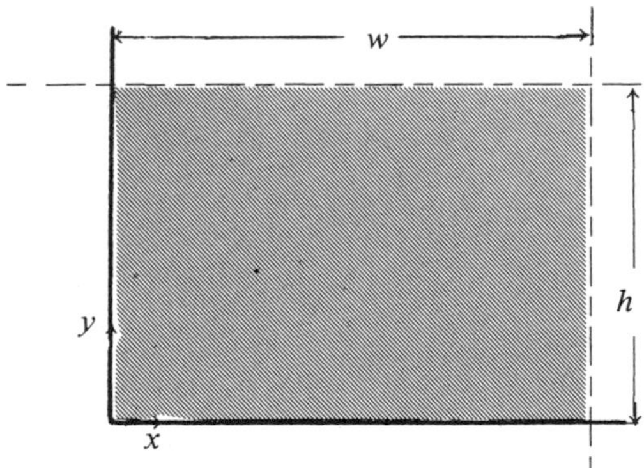

Fig 2.5 Illustration of the solution domain (shaded) for LESSON 5

the time increment DT will be set equal to 50s initially and then increased by a factor of 1.05 at certain time steps, according to a prescription given in the LESSON UPDATE listing (the motive is to preserve accuracy while minimising computing time). The residual-source normalisation factor is here defined in terms of a notional thermal-resistance formula as: $|(T_F - T_C)/(h/k + 1/\alpha)|$, T_C being the temperature at the center of the ingot.

For the requirements of objective (*a*), set $A = 1$ through w and then make RUNs for $\mathrm{Bi} = 10^{-3}$, 1 and 10^2, these values to be specified through α.

Objective (*b*) is achieved by setting $\mathrm{Bi} = 1$ and making additional RUNs with $A = 2$ and 100.

3.5.4 *Programming instructions*

This LESSON follows on very closely from the Base Case, the principle change being the insertion of the convectively cooled boundaries and the symmetry planes: these and other required alterations are summarised in table 2.12, the associated new FORTRAN variables being defined in table 2.11; and UPDATE listings are provided for effecting the changes and making the RUNs.

TABLE 2.11 New FORTRAN variables for LESSON 5

Variable	Quantity	Significance
ALPHA	α	Heat transfer coefficient
BIOT	Bi	Biot number
TF	T_F	Fluid temperature
TI	T_I	Initial temperature of the ingot

3.5.5 *Analysis of results*

(*i*) Plot on graph * 2.11 the variation of T_C^* with t°, T_C^* being the dimensionless temperature at the centre of the ingot, for $A = 1$ and all Biot numbers. Explain the way the behaviour of the temperature depends on the value of Bi.

(*ii*) Compare the temperature decay exhibited in graph 2.11 for small Bi with the analytical solution given by eqn. (2.10). What conclusions may be drawn?

(*iii*) Insert the predictions of T_C^* for $\mathrm{Bi} = 1$ and various A on graph 2.12. Compare the times required for the temperature difference to fall to one-half the initial value and explain the relationship.

(*iv*) Sketch, first on the computer outputs, and then on graphs 2.13 (*a*)–(*c*), isotherms for 3 or 4 selected temperatures at corresponding times for the different aspect ratios and explain the differences in behaviour.

* Graphs will be found on pages 127–135

TABLE 2.12 Programming instructions for LESSON 5

Subroutine	Chapter	Changes
CONTRO	0	◆ Insert ALPHA and TF in COMMON block
		◆ Set heading for exact solution
	1	◆ Set NI, NJ, W and H
		◆ Set material properties, MAXSTP, SORMAX and DT
		◆ Insert TF and ALPHA, and calculate BIOT
	2	◆ Initialise temperature field and set SNORM
	3	◆ Increase time step and calculate boundary temperatures
	4	◆ Print steady-state solution
		◆ Provide formats for output
PROMOD	0	◆ Insert COMMON block as in CONTRO
	1	◆ Make North and East boundaries planes of symmetry
		◆ Apply convective boundary conditions at South and West boundaries

PROBLEM 2 LESSON 5 UPDATE for effecting the changes specified in table 2.12

```
*IDENT,P2L5
*I,CONTRO.37
     1/P2L5/ALPHA,TF
*D,CONTRO.43
     1      /'          ','EXACT ','SOLUTI','ON    ',2*'        '/
*D,CONTRO.57,60
      NI=8
      NJ=8
      W=0.1
      H=0.1
*D,CONTRO.84,85
      IMON=4
      JMON=4
*D,CONTRO.97
      SORMAX=0.04
*D,CONTRO.87,89
      TCON=52.
      CV=460.
      DENSIT=7850.
*D,CONTRO.111,120
      TI=1000.
      DO 2001 I=1,NI
      DO 2001 J=1,NJ
 2001 T(I,J)=TI
*D,CONTRO.92
      MAXSTP=260
```

```
*D,CONTRO.94
      NSTPRI=20
*D,CONTRO.98
      DT=50.
C-----FLUID TEMPERATURE AND HEAT TRANSFER COEFFICIENT
      TF=400.
      ALPHA=.52
      BIOT=ALPHA*H/TCON
      IF (BIOT.LT.0.1) MAXSTP=200
      RATT1=TCON/ALPHA/Y(2)
      RATT2=TCON/ALPHA/X(2)
*D,CONTRO.124,127
      SNORM=ABS((TF-T(NI,NJ))/(H/TCON+0.1/ALPHA))
*D,CONTRO.129
      WRITE(6,2900) H,W,CV,TCON,DENSIT,DT,TI,TF,ALPHA,BIOT,SNORM,NI,NJ
*I,CONTRO.137
      SNORM=ABS((TF-T(NI,NJ))/(H/TCON+0.1/ALPHA))
C-----INCREASE TIME STEP
      IF (NSTEP.LT.140.OR.NSTEP.GT.200.AND.NSTEP.GT.1) DT=DT*1.05
*D,CONTRO.148,149
C-----SET BOUNDARY TEMPERATURES
      DO 3201 I=1,NI
      T(I,1)=(RATT1*T(I,2)+TF)/(RATT1+1.)
 3201 T(I,NJ)=T(I,NJM1)
      DO 3202 J=1,NJ
      T(1,J)=(RATT2*T(2,J)+TF)/(RATT2+1.)
 3202 T(NI,J)=T(NIM1,J)
*I,CONTRO.181
C-----CALCULATION OF EXACT STEADY STATE SOLUTION
      DO 4101 I=1,NI
      DO 4101 J=1,NJ
 4101 T(I,J)=TF
      CALL PRINT (1,1,NI,NJ,IT,JT,X,Y,T,HEDS)
*I,PROBLEM2.1

      1/16X,48H LESSON 5      COOLING OF A RECTANGULAR STEEL BAR
      1/30X,41HWITH PRESCRIBED HEAT TRANSFER COEFFICIENT//
*I,CONTRO.191
      1/16X,40H INITIAL TEMPERATURE OF SOLID, TI -----=,G10.3,2H K
      1/16X,40H FLUID TEMPERATURE, TF ----------------=,G10.3,2H K
      1/16X,40H HEAT TRANSFER COEFFICIENT, ALPHA -----=,G10.3,9H W/M**2 K
      1/16X,40H BIOT NUMBER, BIOT --------------------=,G10.3
*I,PROMOD.15
      1/P2L5/ALPHA,TF
*D,PROMOD.24,28
*D,PROMOD.31,32
      RT=1./(ALPHA*SEW(IL))+1./DS
      SU(2)=SU(2)+TF/RT
      SP(2)=SP(2)-1./RT
*D,PROMOD.38,40
*D,PROMOD.47,48
      RT=1./(ALPHA*SNS(J))+1./DW
      SU(J)=SU(J)+TF/RT
      SP(J)=SP(J)-1./RT
```

An UPDATE of the following form will be required to perform the RUNs of LESSON 5.

```
*IDENT,P2L5R1
*D,P2L5.5
      W=1.Ø
*D,P2L5.15
      DT=.Ø1
*D,P2L5.18
      ALPHA=52Ø.
```

4. References

1. H.S.Carslaw & J.C.Jaeger, *Conduction of Heat in Solids,* Oxford University Press, 1957.
2. V. Arpaci, *Conduction Heat Transfer,* Addison Wesley, 1966.
3. P.J.Schneider, *Conduction Heat Transfer,* Addison Wesley 1955.

Graph for LESSON 1

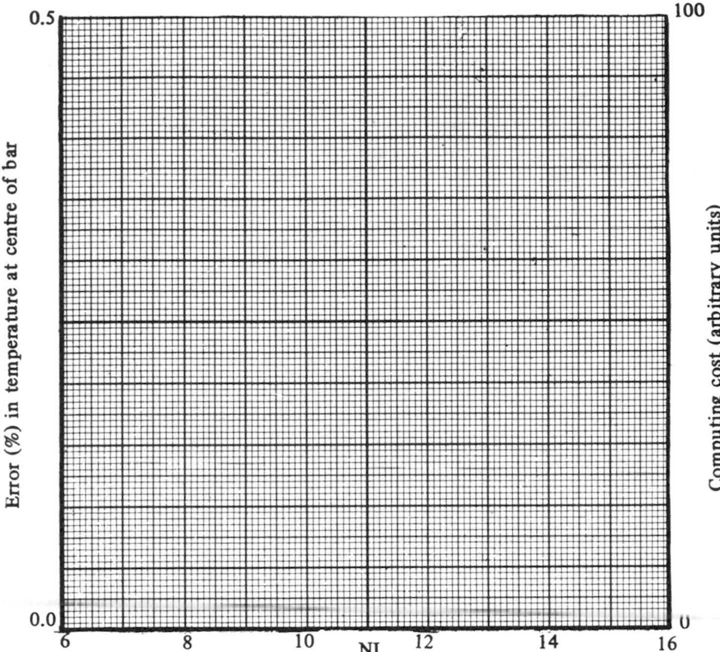

Graph 2.1 Variation of accuracy and computing cost with number of grid intervals

Graphs for LESSON 2

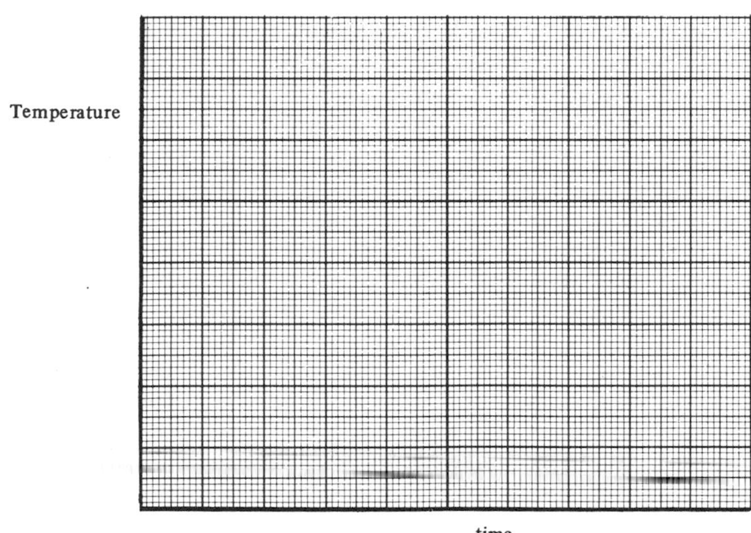

Graph 2.2 Variation of monitoring temperature with time for case of constant conductivity

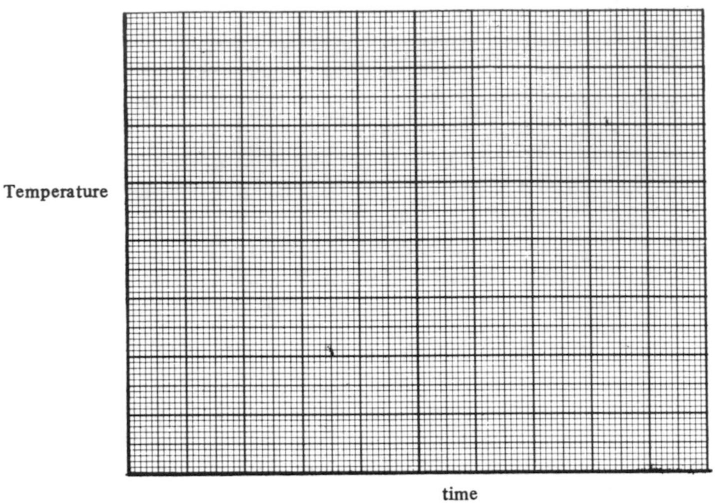

Graph 2.3
Variation of monitoring temperature with time for case of temperature-dependent conductivity

Graphs for LESSON 3

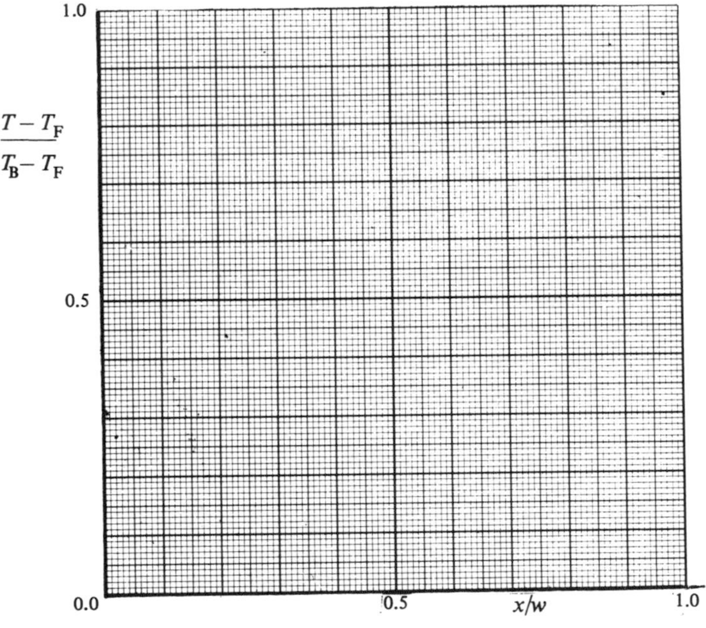

Graph 2.4(*a*)
Temperature at symmetry plane for aspect ratio of 5:1 against distance, with Bi as parameter

Note that the proportions of the plots do not correspond to the aspect ratios

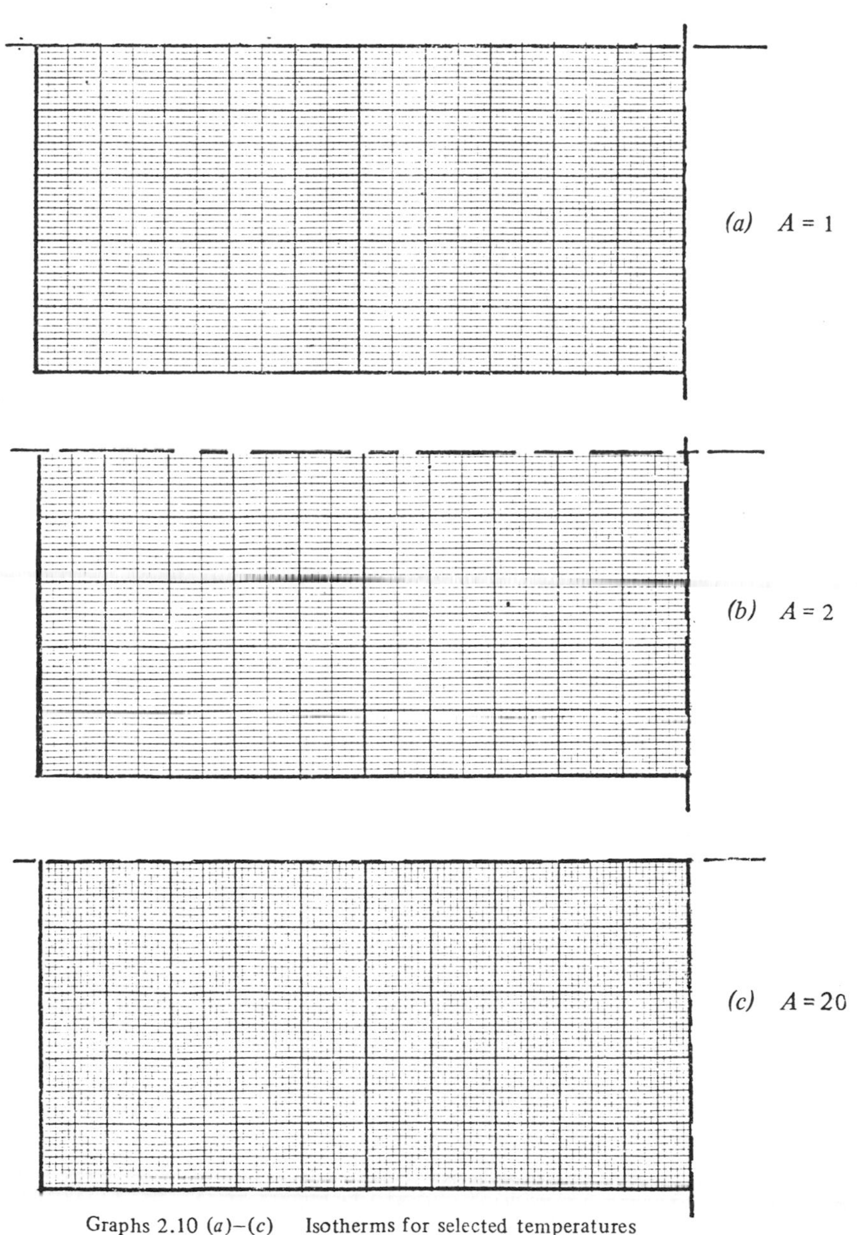

Graphs 2.10 (*a*)–(*c*) Isotherms for selected temperatures at corresponding times, for the three aspect ratios

Graphs for LESSON 5

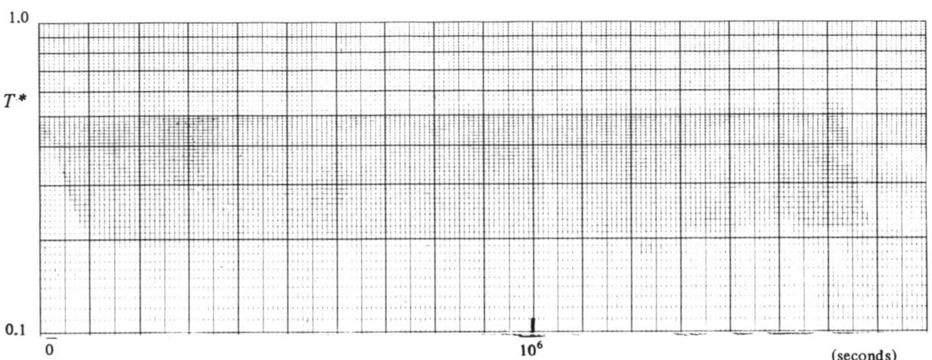

Graph 2.11 Variation of normalised centre temperature with time

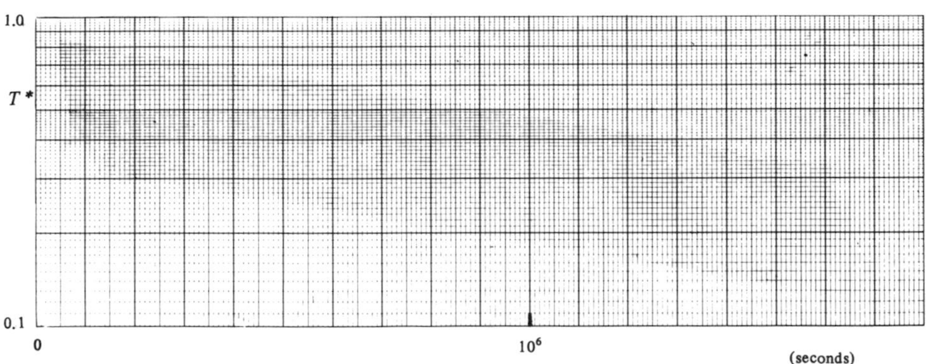

Graph 2.12 Predicted normalised centre temperature for Bi=1 and various aspect ratios

Note that the proportions of the plots do not correspond to the aspect ratios

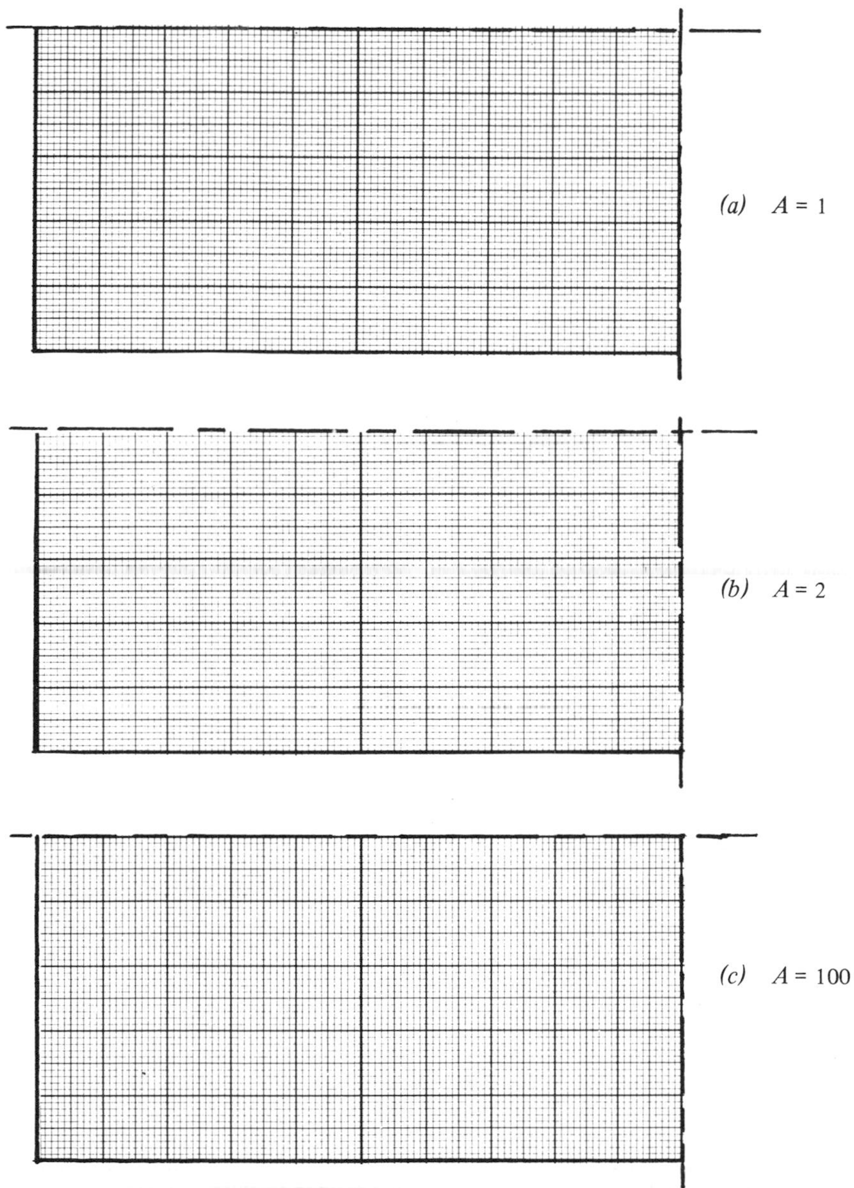

Graphs 2.13 (*a*)–(*c*) Isotherms for selected temperatures at corresponding times, for the three aspect ratios

Applications of TEACH–C
PROBLEM 3
Potential Flow

1. Introduction

In certain types of fluid flow the effects of viscous shear are negligible: examples are

- the air motion in all but the immediate vicinity of an aircraft wing, or
- the impingement of a uniform wind on the side of a building.

This PROBLEM is concerned with analysing impinging jets and an idealised version of a hovercraft using an inviscid model and then examining the results for plausibility.

The omission of the viscous terms greatly simplifies the task of solving the equations of motion, which for a uniform-density, axisymmetric two-dimensional non-viscous flow may be written:*

$$\frac{\partial u}{\partial t} + u\frac{\partial u}{\partial x} + v\frac{\partial u}{\partial r} = -\frac{1}{\rho}\frac{\partial p}{\partial x} \qquad (3.1)$$

$$\frac{\partial v}{\partial t} + u\frac{\partial v}{\partial x} + v\frac{\partial v}{\partial r} = -\frac{1}{\rho}\frac{\partial p}{\partial r} \qquad (3.2)$$

$$\frac{\partial ru}{\partial x} + \frac{\partial rv}{\partial r} = 0 \qquad (3.3)$$

As usual, plane two dimensional flow is recovered by replacing r by 1, and ∂r by ∂y.

The simultaneous solution of these three equations would yield the distribution through the flow of the two velocity components u and v and the pressure field. It is helpful to combine the momentum equations so that the pressure is eliminated; this is easily accomplished by subtracting the x-derivative of (3.2) from the r-derivative of (3.1). The result may be expressed:

* see, e.g., reference [3.1]

$$\frac{D\omega}{Dt} \equiv \frac{\partial r\omega}{\partial t} + \frac{\partial}{\partial x}(ru\omega) + \frac{\partial}{\partial r}(rv\omega) = 0 \tag{3.4}$$

where the quantity ω defined by

$$\omega \equiv \frac{\partial u}{\partial r} - \frac{\partial v}{\partial x} \tag{3.5}$$

is termed the *vorticity* of the fluid. Eqn. (3.4) thus states that the vorticity of any fluid element remains constant. Consider, for example, a flow started from rest. Initially, its vorticity is zero everywhere; according to eqn. (3.4) it will therefore remain zero at all subsequent times, if the boundary values of ω are zero everywhere.

Inviscid flows in which the vorticity is everywhere zero are termed *irrotational:* it is these which we shall concentrate on in this PROBLEM. For such flows the velocity field is determined from the solution of the very simple pair of partial differential eqns. given by the continuity eqn. (3.3) and the requirement that $\omega = 0$, i.e.

$$\frac{\partial u}{\partial r} - \frac{\partial v}{\partial x} = 0 \tag{3.6}$$

Let us now introduce a two-dimensional scalar function called the *stream function* $\psi\ (x, y)$ defined such that

$$\frac{\partial \psi}{\partial r} = \rho r u \qquad \frac{\partial \psi}{\partial x} = -\rho r v \tag{3.7}$$

Lines of constant ψ are *streamlines,* the spacing between which is inversely proportional to the local velocity as may be seen from eqns. (3.7). Moreover the velocities associated with the $\psi\ (x,\ r)$ field satisfy the continuity eqn. (3.3), which may be confirmed by substituting eqns. (3.7) into (3.3).

Substitution of eqns. (3.7) into the irrotationality eqn. (3.6) leads to the important result that

$$\frac{\partial}{\partial x}\left(\frac{1}{r}\frac{\partial \psi}{\partial x}\right) + \frac{\partial}{\partial r}\left(\frac{1}{r}\frac{\partial \psi}{\partial r}\right) = 0 \tag{3.8a}$$

or, for plane flow,

$$\frac{\partial^2 \psi}{\partial x^2} + \frac{\partial^2 \psi}{\partial y^2} = 0 \tag{3.8b}$$

Comparison of these equations with the general form of equation solved by TEACH–C, viz.

$$\rho c_v r \frac{\partial T}{\partial t} - \frac{\partial}{\partial x}\left(rk \frac{\partial T}{\partial x}\right) - \frac{\partial}{\partial y}\left(rk \frac{\partial T}{\partial y}\right) - rs = 0$$

shows that they fit within the framework of the latter provided that ψ is identified with T, $\partial T/\partial t$ and s are both set to zero, and k is set equal to $1/r^2$. It therefore follows that TEACH–C can be used to calculate irrotational flows.

A further important property of irrotational flows may be inferred by multiplying (3.1) by u and adding it to (3.2) multiplied by v. For steady flow the result may be expressed as

$$\frac{\partial}{\partial x}(rup_t) + \frac{\partial}{\partial r}(rvp_t) = 0 \qquad (3.9)$$

where p_t stands for the local "total" pressure of the fluid i.e.

$$p_t \equiv p + \rho(u^2/2 + v^2/2) \qquad (3.10)$$

Eqn. (3.10) has the general solution: p_t = constant along any streamline or vortex line [see ref. 3.1], and hence in the present case constant everywhere in the fluid. Eqn. (3.10) is a form of the well-known *Bernoulli's equation.*

Although there is an extensive body of analytical solutions to eqn. (3.8) for both plane and axisymmetric flows (see for example the texts by Lamb [ref. (3.2)] and Milne-Thomson [ref. (3.3)], there are many situations for which a numerical solving procedure like that of TEACH–C is the easiest, if not the only, route to a solution.

2. Programming

Instructions are given below on how to adapt TEACH–C to the solution of eqns (3.8a) or (3.8b) above for the various circumstances described above. We first establish a basic situation or 'Base Case' on which the individual LESSONs can be regarded as variations. The base case is that of the axisymmetric impingement of fluid against a vertical wall, as illustrated in fig. 3.1. Calculations are performed in the rectangular region bounded by the wall, the symmetry axis and arbitrarily-placed planes within the flow. A uniform horizontal velocity u is prescribed across the vertical plane, and a uniform radial velocity v is set along the horizontal plane, these specifications leading, via eqns. (3.7), to the boundary prescriptions on ψ indicated in the figure (there is of course no flow through the wall or symmetry axis).

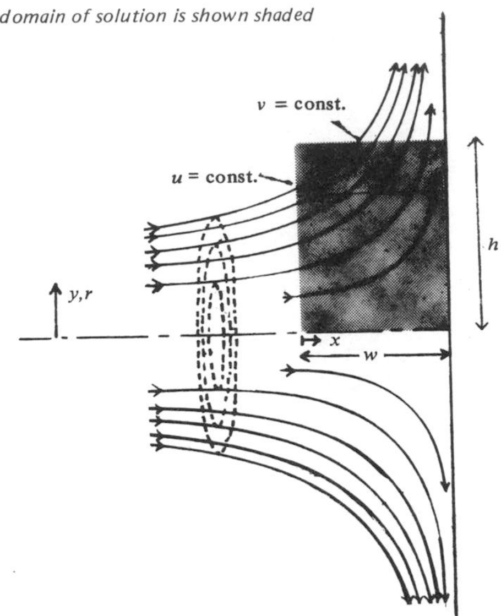

Figure 3.1 Axisymmetric impinging flow

For radial 'conduction' between adjacent cells, the effective diffusivity is taken as that evaluated at the inter-cell boundary, rather than as the arithmetic mean of that at the two nodes. (It is readily shown that if the ψ variation is purely radial this choice coupled with the presumed linear variation of ψ between nodes leads to exactly the correct diffusional flux of ψ.) For *axial* conduction the mean effective axial diffusivity (k_{ax}, say) is obtained simply by integrating over the cell face:

$$r_p k_{ax} \delta r_{ns} = \int_{r_s}^{r_n} r \cdot \frac{1}{r^2} \, dr \qquad (3.11)$$

or

$$k_{ax} = \ell n \, (r_n / r_s)/(r_p \, \delta r_{ns})$$

From equation (3.7) it is clear that $\psi \sim r^2$ near the axis. The value of T(I,2) is therefore set, with the help of GREAT, to the value corresponding to this parabolic variation.

Solution of (3.8) provides the stream function field. In practice the quantities of interest are the velocities, u and v, and the pressure p. Arrangements are therefore made in CONTRO to calculate these quantities once a converged solution has been obtained for the stream function. This is very easy to do; given the values of ψ we can use finite-difference versions of the eqns. (3.7) to generate the values of u and v. Then, using these, we apply eqn. (3.10) to obtain values of p, imposing the condition that p_t be uniform over the flow field.

The modifications to TEACH–C required to simulate this flow are outlined in table 3.1 below; no new FORTRAN variables are required to do this.

Table 3.1 is followed by a listing of the UPDATE required to effect the changes specified in the table.

TABLE 3.1 Programming instructions for PROBLEM 3 Base Case

Subroutine	Chapter	Change
CONTRO	0	♦ Provide additional headings for output
	1	♦ Set material properties to unity ♦ Set SORMAX and DT, and INTIME=.FALSE.
	2	♦ Assign boundary values for ψ ♦ Set SNORM
	4	♦ Insert sequence to compute and print u, v and p using the arrays TOLD(I,J) and GAMH(I,J) ♦ Supply PROBLEM heading
PROPS	1	♦ In axisymmetric case, compute effective diffusion coefficient using logarithmic-mean averaging.
CALCT	3	♦ Suppress contribution to RESORT of spurious residual sources in cells adjacent to axis
PROMOD	1	♦ Modify north boundary diffusion coefficient
	2	♦ For axisymmetric case evaluate T(I,2) from $\psi \sim r^2$ near axis

PROBLEM 3 Base Case UPDATE for effecting the changes specified in table 3.1

```
*IDENT,PROBLEM3
*D,CONTRO.40,41
      CHARACTER*6 HEDU(6),HEDV(6),HEDP(6)
      DATA HEDT(1),HEDT(2),HEDT(3),HEDT(4),HEDT(5),HEDT(6)
     1     /'      ','    ST','REAM F','UNCTIO','N     ','      '/
      DATA HEDU(1),HEDU(2),HEDU(3),HEDU(4),HEDU(5),HEDU(6)
     1     /'      ','   U V','ELOCIT','Y     ',2*'      '/
      DATA HEDV(1),HEDV(2),HEDV(3),HEDV(4),HEDV(5),HEDV(6)
     1     /'      ','   V V','ELOCIT','Y     ',2*'      '/
      DATA HEDP(1),HEDP(2),HEDP(3),HEDP(4),HEDP(5),HEDP(6)
     1     /2*'      ','PRESSU','RE    ',2*'      '/
*D,CONTRO.87,89
      TCON=1.0
      CV=1.0
      DENSIT=1.0
*D,CONTRO.97,98
      SORMAX=0.001
      DT=0.0
*D,CONTRO.100
      INTIME=.FALSE.
*D,CONTRO.111
      TTOP=1.0
*D,CONTRO.113
      TLEFT=1.0
*I,CONTRO.114
      H2=H
      IF (INCYLY) H2=H*H
      H2W=H2*W
      T(1,NJ)=TTOP
*D,CONTRO.116
      T(I,NJ)=TTOP*(W-X(I))/W
*D,CONTRO.119
      T(1,J)=TLEFT*RY(J)*Y(J)/H2
*D,CONTRO.124,127
      SNORM=ABS(TRIGHT-TLEFT)*(H/W+W/H)
*D,CONTRO.129
      WRITE (6,2900) H,W,SNORM,NI,NJ
*I,CONTRO.181
C-----OBTAIN PREDICTED VELOCITY FIELDS AND PRESSURE
C-----PREDICTED U VELOCITY
      NJM2=NJ-2
      DO 4120 I=1,NI
      GAMH(I,2)=(T(I,1)*(Y(2)-Y(3))**2+T(I,2)*Y(3)*(2.*Y(2)-Y(3))-
     1           T(I,3)*Y(2)**2)/(RY(2)*Y(2)*Y(3)*(Y(2)-Y(3)))
      GAMH(I,1)=GAMH(I,2)
      DO 4110 J=3,NJM2
      GAMH(I,J)=(T(I,J+1)-T(I,J-1))/((Y(J+1)-Y(J-1))*RY(J))
 4110 CONTINUE
      GAMH(I,NJM1)=(T(I,NJ)-T(I,NJM2))/((YV(NJ)-Y(NJM2))*RY(NJM1))
      GAMH(I,NJ)=(T(I,NJ)-T(I,NJM1))/((YV(NJ)-Y(NJM1))*RV(NJ))
 4120 CONTINUE
      CALL PRINT(1,1,NI,NJ,IT,JT,X,Y,GAMH,HEDU)
C-----STORE U CONTRIBUTION TO PRESSURE IN TOLD
      DO 4130 I=1,NI
      DO 4130 J=1,NJ
      TOLD(I,J)=GAMH(I,J)**2
 4130 CONTINUE
C-----PREDICTED V VELOCITY
      DO 4140 I=1,NI
 4140 GAMH(I,1)=0.0
      NIM2=NI-2
      DO 4170 J=2,NJ
      RAD=RY(J)
      IF (J.EQ.NJ) RAD=RV(J)
      DO 4150 I=1,2
 4150 GAMH(I,J)=(T(1,J)-T(I+1,J))/(X(I+1)*RAD)
      DO 4160 I=3,NIM2
      GAMH(I,J)=(T(I-1,J)-T(I+1,J))/((X(I+1)-X(I-1))*RAD)
 4160 CONTINUE
      GAMH(NIM1,J)=(T(NIM2,J)-T(NI,J))/((XU(NI)-X(NIM2))*RAD)
      GAMH(NI,J)=(T(NIM1,J)-T(NI,J))/((XU(NI)-X(NIM1))*RAD)
 4170 CONTINUE
```

```
      CALL PRINT(1,1,NI,NJ,IT,JT,X,Y,GAMH,HEDV)
C-----STORE PRESSURE IN TOLD
      PREF=DENSIT*0.5*(TOLD(NI,1)+GAMH(NI,1)**2)
      DO 4180 I=1,NI
      DO 4180 J=1,NJ
      TOLD(I,J)=PREF-DENSIT*0.5*(TOLD(I,J)+GAMH(I,J)**2)
 4180 CONTINUE
C-----PRINT PRESSURE
      CALL PRINT(1,1,NI,NJ,IT,JT,X,Y,TOLD,HEDP)
*D,CONTRO.184,185
 2900 FORMAT(1H1,16X,42H PROBLEM 3    IRROTATIONAL STAGNATION FLOW//
*D,CONTRO.188,191
*I,PROPS.12
     1/CONVAR/IT,JT,INTIME,DT,RESORT,URFT,GREAT
      LOGICAL INCYLY
*I,PROPS.15
      IF (INCYLY) GO TO 1110
*I,PROPS.20
C-----INTERPRET THERMAL CONDUCTIVITY AS TCON/(R*R1) IN AXISYMMETRIC CASE
C-----1/R1 IS A ;MEAN; VALUE FOR 1/R
 1110 DO 1120 I=1,NI
      GAMH(I,NJ)=TCON
      DO 1120 J=1,2
      GAMH(I,J)=GREAT
 1120 CONTINUE
      DO 1130 J=3,NJM1
      GAM=TCON*ALOG(RV(J+1)/RV(J))/(RY(J)*SNS(J))
      DO 1130 I=1,NI
      GAMH(I,J)=GAM
 1130 CONTINUE
      RETURN
*D, CALCT.16
      LOGICAL INTIME,INCYLY
*I,CALCT.62
      IF (INCYLY.AND.J.EQ.2) RESOR=0.0
*I,PROMOD.15
      LOGICAL INCYLY
*D,PROMOD.25
      DN=TCON*SEW(IL)*RDYN/RV(NJ)**2
*D,PROMOD.50
C-----CALCULATE MORE ACCURATE THERMAL CONDUCTIVITIES AT INTERNAL
C-----NORTH AND SOUTH CELL FACES FOR AXISYMMETRIC CASE
 1400 IF (.NOT.INCYLY) RETURN
      NJM2=NJ-2
      DO 1410 J=2,NJM2
      DN=TCON*SEW(IL)/(RV(J+1)*DYNP(J))
      AN(J)=DN
      AS(J+1)=DN
 1410 CONTINUE
C --- FIX STREAM FUNCTION TO ZERO AT AXIS.
C --- ASSUME PSI VARIES AS Y**2 TO OBTAIN T(I,2)
      SP(2)=SP(2)-GREAT
      SU(2)=SU(2)+T(IL,3)*(Y(2)/Y(3))**2*GREAT
      RETURN
```

3. LESSONs

3.1 LESSON 1 Plane and axisymmetric stagnation flow

3.1.1 *Physical background*

There are many flows of practical interest where a fluid stream is directed nearly at right-angles to a wall and where, due to the pressure field thus created, the flow is deflected in a direction parallel to the wall. One example is the flow issuing from the engines of a vertical take-off aeroplane as it leaves the ground.

For the case where fluid enters the region of interest with uniform u velocity and zero vorticity an irrotational stagnation flow is created, exactly as in the Base Case. For this situation there are simple analytic solutions [see, e.g., Ref (3.4)] for the stream function distribution. For plane flow,

$$\psi = \rho y (w - x)/wh \tag{3.11}$$

from which with the aid of eqns. (3.7)

$$\begin{aligned} u &= (w - x)/wh \\ v &= y/wh \end{aligned} \tag{3.12}$$

Correspondingly, for an axisymmetric flow

$$\psi = \rho r^2 (w - x)/wh^2 \tag{3.13}$$

which leads to

$$\begin{aligned} u &= 2 (w - x)/(w h^2) \\ v &= r/(w h^2) \end{aligned} \tag{3.14}$$

where the notation is that of Fig. (3.1)

In both cases the local pressure is obtained from the fact that the total pressure is uniform throughout the flow:

$$p = p_t - \frac{\rho}{2} (u^2 + v^2) \tag{3.10}$$

3.1.2 *Objective*

To obtain numerical solutions of the problems of plane and axisymmetric stagnation flow using various computational grids, and to determine how well these solutions agree with the analytical solutions to the same problems.

3.1.3 *Method*

We first make calculations for plane flow, setting NI=**NJ**=12, and then altering **NJ** first to 8 and then to 16. We next repeat this process for axisymmetric flow.

3.1.4 *Programming*

The only changes of significance relate to the inclusion of sequences for calculating and printing the exact solution when a converged numerical solution has been obtained. The spare arrays TOLD (I, J) and GAMH (I, J) are repeatedly used as temporary storage for different elements of the exact solution prior to printout.

Table 3.2 summarises the programming changes required for LESSON 1. No new FORTRAN variables are required. Table 3.2 is followed by a listing of the UPDATE which effects these changes.

TABLE 3.2 Programming instructions for LESSON 1

Subroutine	Chapter	Change
CONTRO	0	♦ Provide heading for exact solution
	4	♦ Compute and print exact solutions for ψ, u and v
		♦ Supply LESSON heading

LESSON 1 UPDATE for effecting the changes specified in table 3.2

```
*IDENT,P3L1
*D,CONTRO.43
     1     /'          ','EXACT ','SOLUTI','ON    ',2*'      '/
*D,CONTRO.56
      INCYLY=.TRUE.
*D,CONTRO.58
      NJ=20
*I,CONTRO.181
C-----CALCULATE EXACT STREAM FUNCTION DISTRIBUTION
      DO 4101 J=2,NJM1
 4101 TOLD(1,J)=TLEFT*RY(J)*Y(J)/H2
      TOLD(1,NJ)=TLEFT
      DO 4103 I=2,NIM1
      DO 4102 J=2,NJM1
      TOLD(I,J)=TTOP*RY(J)*Y(J)*(W-X(I))/H2W
 4102 CONTINUE
      TOLD(I,NJ)=TTOP*(W-X(I))/W
 4103 CONTINUE
      CALL PRINT(1,1,NI,NJ,IT,JT,X,Y,TOLD,HEDS)
      WRITE(6,4901)
*I,PROBLEM3.44
C-----EXACT U VELOCITY
      TLH2W=TLEFT/H2W
      IF (INCYLY) TLH2W=TLH2W*2.0
      DO 4131 J=1,NJ
      DO 4131 I=1,NI
      GAMH(I,J)=TLH2W*(W-X(I))
 4131 CONTINUE
      CALL PRINT(1,1,NI,NJ,IT,JT,X,Y,GAMH,HEDS)
      WRITE(6,4901)
*I,PROBLEM3.66
C-----EXACT V VELOCITY
      TTH2W=TTOP/H2W
      DO 4181 I=1,NI
      DO 4181 J=1,NJ
      GAMH(I,J)=TTH2W*Y(J)
 4181 CONTINUE
      CALL PRINT(1,1,NI,NJ,IT,JT,X,Y,GAMH,HEDS)
      WRITE(6,4901)
*I,PROBLEM3.69
     1/16X,52H LESSON 1      PLANE AND AXISYMMETRIC STAGNATION FLOW//
*I,CONTRO.201
 4901 FORMAT(///)
```

To perform the RUNs of LESSON 1, students will require an UPDATE of the following kind:

```
*IDENT, P3L1R1
*D, CONTRO.56
      INCYLY  = .TRUE.
*D, CONTRO.58
      NJ = 4
*D, CONTRO.85
      JMON = 2
```

3.1.5 *Analysis of results*

(*i*) For the finest grid used, sketch directly on to the computer output, and then on graphs 3.1 and 3.2 a few representative lines of constant stream-function (i.e. streamlines) for the plane and axisymmetric cases respectively. Are these flow patterns consistent with your expectations? Identify and explain the differences between the two cases.

(*ii*) Sketch contours of constant pressure (*isobars*) for the plane and axisymmetric cases on graphs 3.3 and 3.4. What shape are the lines in each case? Verify your conclusions by reference to eqn. (3.10).

(*iii*) Plot on graph 3.5 the variation of $p_t - p$ along the wall surface with x for both cases. Explain the slopes of the curves – are they consistent with eqn. (3.10)?

(*iv*) Next determine the influence of the grid by plotting on graph 3.6 the variation of the percentage difference between the exact and computed values of ψ at the central node of the grid as a function of NJ for both cases. Explain the cause of the variation and the differences, if any, between the curves.

(*v*) Plot, again on graph 3.6, the variation with NJ of the number of iterations required for convergence and explain what practical conclusions may be drawn from the complete set of results shown on this graph.

3.2 LESSON 2 The flow pattern created beneath a stationary hovercraft

3.2.1 *Physical background*

The flow configuration examined in this LESSON is shown in fig. 3.2 which represents an idealisation of that created beneath an "air-cushion" vehicle or hovercraft.

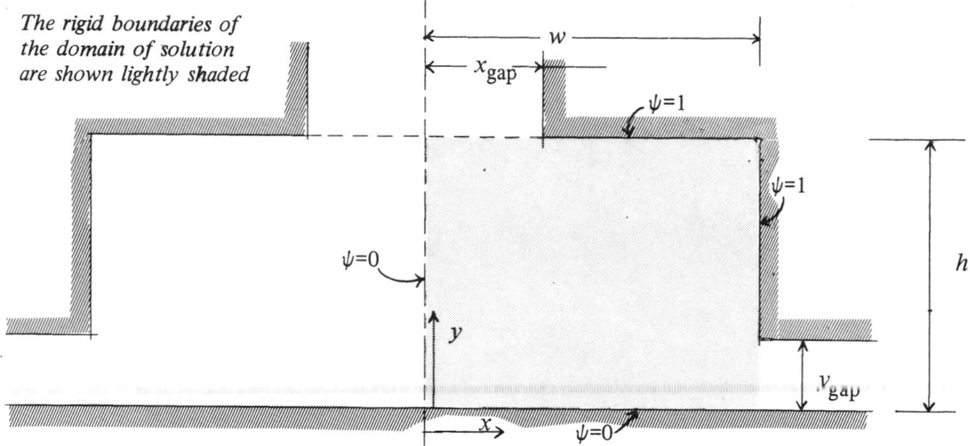

Fig. 3.2 Illustration of a stationary hovercraft

The ground coincides with the plane $y = 0$, and the hovercraft is presumed to extend sufficiently far in the direction normal to the paper that we may treat the flow as two-dimensional. Air and combustion products from the turbines of the vessel are admitted through the slot in the roof of the support structure and escape between the lower edge of the "skirts" and the ground. The ratio $w:h$ is here kept fixed at 5:1 which is fairly representative of practical geometries.

A practical limitation of the use of hovercraft is the unevenness of the terrain or water over which the craft must travel. The more uneven the surface the larger on average, will be the gap beneath the hovercraft skirts. As a result, the amount of air required to "cushion" the craft is greater. There is of course a limit to the air supply that the craft's turbines can provide and so the hovercraft is, for example, unable to operate in very rough seas.

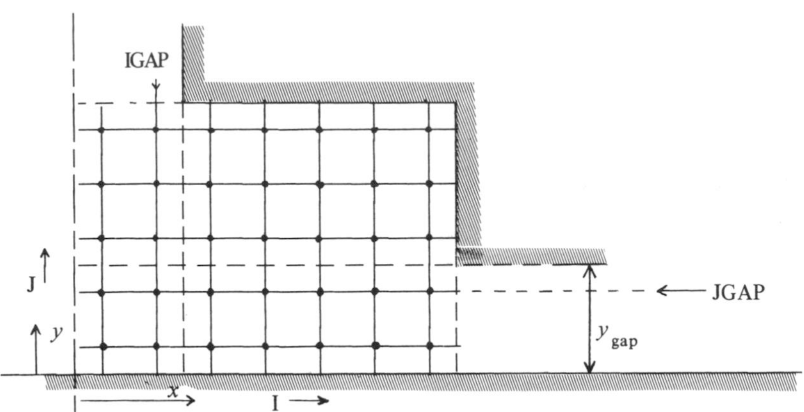

Fig. 3.3 The grid and associated notation for the hovercraft calculations

3.2.2 *Objective*

To examine, with the irrotational flow model, the flow pattern created beneath an air cushion vehicle and to determine the dependence of the pressure distribution on the relative height of the skirt and the mass flow rate.

3.2.3 *Method*

Perform calculations for the case of uniform inlet and outlet velocities. Firstly, make three RUNs varying the skirt gap y_{gap} from $0.1\,h$ to $0.5\,h$. Then, for the two largest gaps make further RUNs adjusting the inlet velocities until the average pressure along the underside of the hovercraft floor is approximately the same as in the RUN for the smallest skirt height. (It is suggested that, for the latter explorations, the inlet velocity be increased by factors of two until the desired pressure level is exceeded: then obtain the required velocity by interpolation).

3.2.4 *Programming*

The symmetry of the situation allows the calculations to be confined to half the hovercraft cross-section, as indicated in fig. 3.2. The boundary values of ψ for this domain are worked out from the prescriptions on the velocities normal to the boundaries, via eqns. (3.7).

A uniformly-spaced grid is employed, as illustrated in fig. 3.3, with NI = NJ = 10. The height of the skirt gap is specified through the index JGAP, the height being (Y(JGAP) + Y(JGAP + 1))/2.

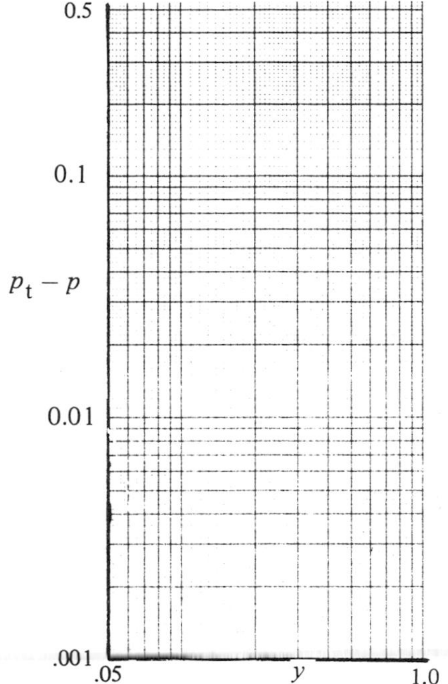

Graph 3.5 Variation of $p_t - p$ along the wall surface (LESSON 1)

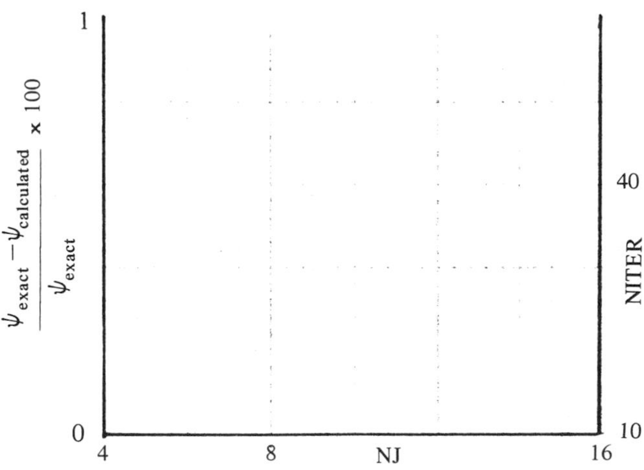

Graph 3.6 Variation of percentage error in stream finction and of number of iterations to convergence, plotted with NJ. (LESSON 1)

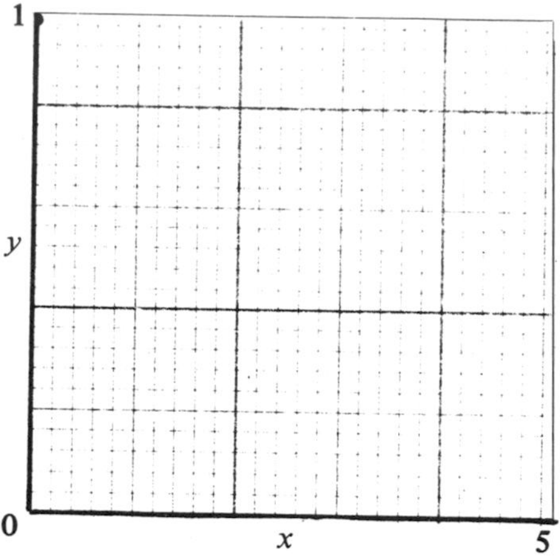

Graph 3.7 Stream function for $y_{gap} = 0.5\ h$ (LESSON 2)

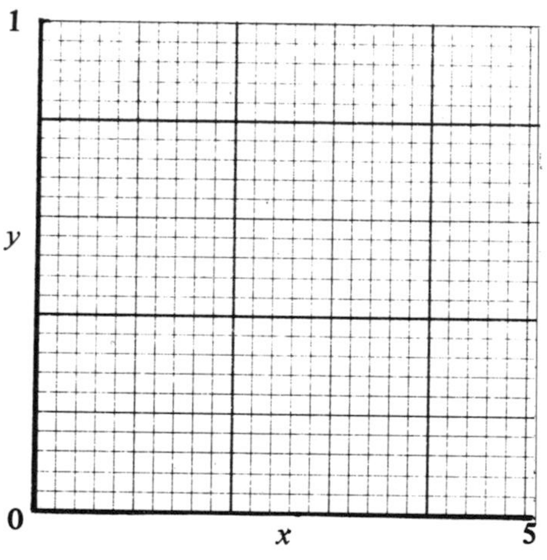

Graph 3.8 Stream function for $y_{gap} = 0.3\ h$ (LESSON 2)

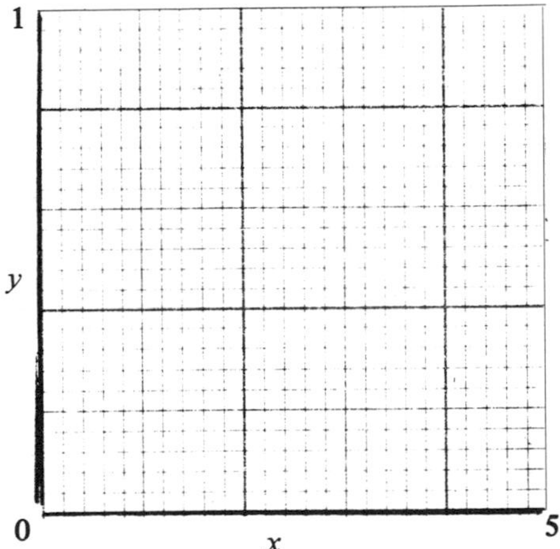

Graph 3.9 Stream function for $y_{gap} = 0.1\ h$ (LESSON 2)

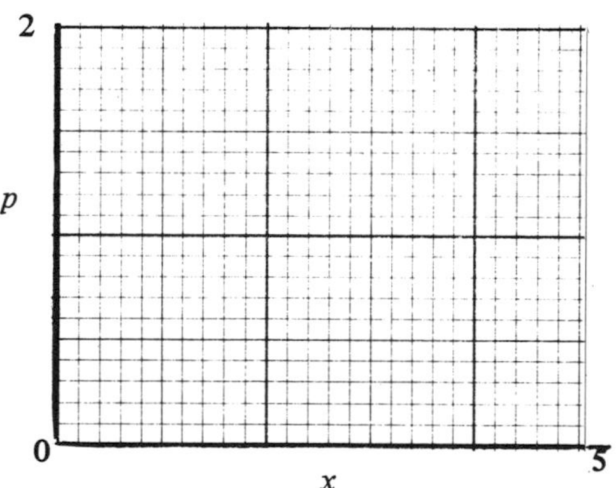

Graph 3.10 Distribution of pressure adjacent to top wall (LESSON

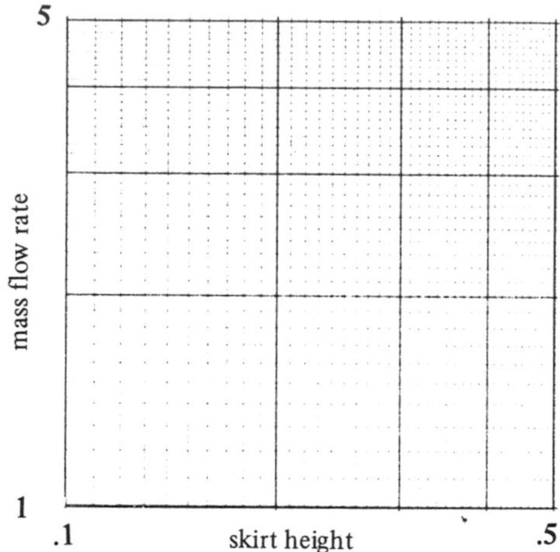

Graph 3.11 Variation of mass flow rate with skirt height for a given mean pressure (LESSON 2)

Applications of TEACH–C
PROBLEM 4
Fully-developed Flow in Non-circular Ducts

1. Introduction

Often in engineering practice a fluid flows through a straight duct of non-circular but effectively uniform cross-section, the length of the duct being very much greater than the maximum section. Examples of such situations include flows in;

- lubrication passages for bearings;
- polymer extrusion plant;
- compact heat-exchangers;
- air-conditioning ducting;

and there are many others. In the first two examples, the flow is almost invariably laminar, while in the latter two it is usually turbulent. This PROBLEM is concerned with the analysis of such situations, albeit with some simplifications.

The "boundary layers" on the walls of the duct, where the flow is significantly retarded by viscous action, thicken with increasing distance from the duct entrance. Eventually, after a certain distance termed the *entry length*, they merge at the duct centre. The velocity distribution is subsequently independent of further distance along the duct, and it is termed *fully-developed.* The magnitude of the entry length depends to some extent on the cross-sectional shape of the duct and on whether the flow is laminar or turbulent. As a guide, it is known that for a pipe of circular cross-section the entry length is about 0.03 x Re x the pipe diameter, where Re is the *pipe Reynolds number,* for laminar flow, and typically about 50 diameters for turbulent flow. (see ref. [4.1]).

Since the velocity distribution for fully-developed flow in a straight duct of constant cross-section is independent of distance along the duct, the problem of determining the distribution in a non-circular section is a two-dimensional one. Furthermore, for laminar flow in straight ducts it can be shown that the velocities in the cross-sectional plane of the duct are zero.* In these circumstances, the conservation of momentum

---

*Excluding cases where part of the duct boundary is moving.

is expressed very simply via the single equation:

$$\frac{\partial}{\partial x}\left(\mu \frac{\partial W}{\partial x}\right) + \frac{\partial}{\partial y}\left(\mu \frac{\partial W}{\partial y}\right) - \frac{dp}{dz} = 0 \qquad (4.1)$$

where x and y are the co-ordinates in the cross-sectional plane of the duct, z is the co-ordinate in the longitudinal direction of the duct, W is the z-direction velocity, p is pressure, and μ is the dynamic viscosity.

Comparison of eqn. (4.1) with the general Cartesian form of equation solved by TEACH–C, viz:

$$\rho c_v \frac{\partial T}{\partial t} - \frac{\partial}{\partial x}\left\{k \frac{\partial T}{\partial x}\right\} - \frac{\partial}{\partial y}\left\{k \frac{\partial T}{\partial y}\right\} - s = 0$$

shows that it fits within the framework of the latter, provided that

- $\partial T/\partial t$ is set to zero
- W is identified with T
- μ ... k
- dp/dz ... s.

It follows that TEACH–C can be used to calculate fully-developed duct flows of considerable complexity.

Eqn. (4.1) may also be recognised as a form of *Poisson's equation*, for which analytical solutions are known to exist for certain simple situations (notably those for which the flow is laminar, the viscosity uniform and the duct of simple cross-section). These solutions, as well as being useful in their own right, may be used to test the accuracy of numerical procedures like that employed in TEACH–C.

2. Programming

Instructions will now be given on how to adapt the TEACH–C code to provide the solution of eqn. (4.1) for fully-developed duct flows of the kinds mentioned above. In preparation for this we shall establish a basic situation or 'Base Case', on which the individual LESSONs may be regarded as variations. The Base Case in question is that of fully-developed uniform-viscosity laminar flow in a rectangular duct of width w_{duct} and height h_{duct}, as illustrated in fig. 4.1, on page 4–3.

In these circumstances, there is symmetry of both the duct and the flow about the horizontal and vertical bisectors of the cross-section. This allows calculations to be confined to a single quadrant, if desired. However, in order to allow for some measure of asymmetry to be introduced later in

The domain of solution is shown shaded

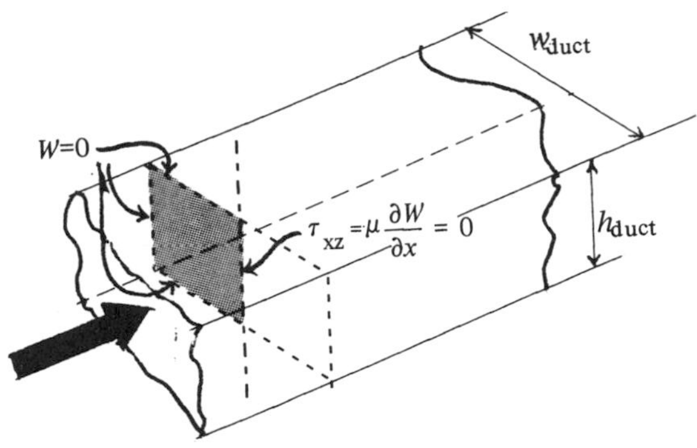

Fig. 4.1 Illustration of PROBLEM 4 Base Case, depicting fully-developed flow in a long rectangular duct.

the LESSONs, just one plane of symmetry will be invoked in the Base Case. The selected solution domain is shown shaded in fig. 4.1; it has dimensions w ($\equiv w_{duct}/2$) and h ($\equiv h_{duct}$). The boundary conditions are

- zero velocity W on the left, top and bottom surfaces
- zero shear-stress τ_{xz} (($\equiv$ $\mu\, \partial W/\partial x$) on the symmetry plane. †

Adaptation of the standard TEACH–C program for this situation entails:

- modifying the headings and titles to reflect the change from heat conduction to fluid flow;
- specifying the grid and flow conditions;
- inserting the (integrated) pressure-gradient source terms and boundary conditions;

---

† Following customary notation the suffix 'z' in τ_{xz} denotes the direction of the stress and 'x' the normal to the surface on which it acts. For an incompressible Newtonian fluid the stresses acting in the z direction are

$$\tau_{yz} = \mu\left(\frac{\partial u}{\partial y} + \frac{\partial v}{\partial x}\right) \qquad \tau_{zy} = \mu\left(\frac{\partial v}{\partial z} + \frac{\partial w}{\partial y}\right) \qquad \tau_{zz} = \mu\left(\frac{\partial w}{\partial x} + \frac{\partial u}{\partial z}\right)$$

In fully-developed flow the z-derivative of all components is zero.

- calculating the following quantities from the converged solutions;
  - the cross-sectional area A of the duct;
  - the mass-flow rate $\dot{m}$;
  - the average velocity $\overline{W} = \dot{m} / \rho A$;
  - the hydraulic diameter $D_h \equiv 4A/P$, where P is the *wetted perimeter*;
  - the Reynolds number
    $\mathrm{Re} \equiv \rho \overline{W} D_h / \mu$
  - the friction factor
    $f \equiv \tau_{wall} / \frac{1}{2} \rho \overline{W}^2$

    where τ_{wall} is the average wall shear-stress.

These and other changes are summarised in table 4.2, which is preceded by a list, in table 4.1, of the new FORTRAN variables required for this PROBLEM. These are followed by a listing of the UPDATE which effects the changes.

Table 4.1 New FORTRAN variables for PROBLEM 4 Base Case

Variable	Quantity	Significance
AREA	A	Cross-sectional area of duct
DH	D_h	Hydraulic diameter
DPDZ	dp/dz	Pressure gradient
FFACT	f	Friction factor
FLOW	$\dot{m}$	Mass-flow rate
HDUCT	h_{duct}	height of duct
RE	Re	Reynolds number
WAVE	$\overline{W}$	Average velocity
WDUCT	w_{duct}	Width of duct

Table 4.2 Programming instructions for PROBLEM 4 Base Case

Subroutine	Chapter	Change
CONTRO	0	♦ Put DPDZ into COMMON ♦ Change heading
	1	♦ Set NI=7 and introduce WDUCT and HDUCT ♦ Set material properties ♦ Select steady-state option
	2	♦ Introduce pressure gradient DPDZ ♦ Calculate residual-source normalisation using DPDZ
	3	♦ Calculate boundary values on symmetry plane
	4	♦ Calculate and print out flow rate, mean velocity, Reynolds number and friction factor ♦ Insert Formats for headings
PROMOD	0	♦ COMMON block as in CONTRO
	1	♦ Introduce momentum source due to pressure gradient ♦ Introduce symmetry boundary on right wall

PROBLEM 4 Base Case UPDATE for effecting the changes specified in Table 4.2:

```
*IDENT,PROBLEM4
*I,CONTRO.37
     1/P4L1/DPDZ,SSN(22),SSS(22),SSE(22),SSW(22)
*D,CONTRO.40,41
      DATA HEDT/'          ','AXIAL ','VELOCI','TY    ',2*'          '/
*D,CONTRO.57
      NI=7
*D,CONTRO.59
      W=0.5
*D,CONTRO.87
      VISCOS=1.0
```

```
*D,CONTRO.89
      DENSIT=1.0
*D,CONTRO.111
      TTOP=0.0
C-----AXIAL PRESSURE GRADIENT
      DPDZ=-100.
*D,CONTRO.100
      INTIME=.FALSE.
*D,CONTRO.97
      SORMAX=0.0001
*D,CONTRO.124,127
      SNORM=-DPDZ*W*H
*D,CONTRO.129
      WRITE(6,2900)H,W,VISCOS,DPDZ,DENSIT,DT,SNORM,NI,NJ
*I,CONTRO.181
C-----CALCULATION OF FLOW RATE,MEAN VELOCITY AND REYNOLDS NUMBER
      FLOW=0.0
      AREA=0.0
      DO 400 I=2,NIM1
      DO 400 J=2,NJM1
      DA=RY(J)*SEW(I)*SNS(J)
      AREA=AREA+DA
      FLOW=FLOW+DENSIT*T(I,J)*DA
  400 CONTINUE
      FLOW=2.*FLOW
      AREA=2.*AREA
      WAVE=FLOW/(DENSIT*AREA)
      DH=4.*AREA/(2.*(H+2.*W))
      RE=DH*WAVE*DENSIT/VISCOS
C-----CALCULATION OF WALL SHEAR STRESS AND FRICTION COEFFICIENT
      DXW=X(2)
      FORCEW=0.0
      DO 401 J=2,NJM1
      SSW(J)=-VISCOS*(T(2,J)-TLEFT)/DXW
  401 FORCEW=FORCEW+SSW(J)*SNS(J)
      DYS=Y(2)
      DYN=YV(NJ)-Y(NJM1)
      FORCES=0.0
      FORCEN=0.0
      DO 402 I=2,NIM1
      SSN(I)=-VISCOS*(T(I,NJM1)-TTOP)/DYN
      FORCEN=FORCEN+SSN(I)*SEW(I)
      SSS(I)=-VISCOS*(T(I,2)-TBOT)/DYS
  402 FORCES=FORCES+SSS(I)*SEW(I)
      FORCET=2.*(FORCEW+FORCEN+FORCES)
      SSAVE=-FORCET/(2.*(H+2.*W))
      FFACT=SSAVE/(0.5*DENSIT*WAVE**2)
      WRITE(6,410)AREA,FLOW,WAVE,DH,RE,FFACT
*D,CONTRO.184,185
 2900 FORMAT(1H1,54X,9HPROBLEM 4//38X,65HFULLY-DEVELOPED FLOW IN DUCTS
     1                                               //
*D,CONTRO.188,189
     2/16X,40H VISCOSITY, VISCOS---------------------=,G10.3
     2/16X,40H PRESSURE GRADIENT, DPDZ---------------=,G10.3
*I,CONTRO.194
  410 FORMAT(1H0,29X,'DUCT AREA=',1PE12.3/30X,'MASS FLOW RATE=',1PE12.3/
     130X,'AVERAGE VELOCITY=',1PE12.3/30X,'HYDRAULIC DIAMETER=',1PE12.3/
     230X,'REYNOLDS NUMBER=',1PE12.3/30X,'FRICTION FACTOR=',1PE12.3)
*I,CALCT.15
     1/P4L1/DPDZ,SSN(22),SSS(22),SSE(22),SSW(22)
*I,PROMOD.15
     1/P4L1/DPDZ,SSN(22),SSS(22),SSE(22),SSW(22)
*D,PROMOD.38,40
*I,PROMOD.18
C-----MOMENTUM SOURCE DUE TO PRESSURE GRADIENT
      DO 1310 J=2,NJM1
      DV=RY(J)*SEW(IL)*SNS(J)
 1310 SU(J)=SU(J)-DPDZ*DV
```

3. LESSONs

3.1 LESSON 1 Laminar flow in rectangular-sectioned ducts

3.1.1 *Physical Background*

The rectangular-sectioned duct shown in fig. 4.1 is the simplest and, in practice, the most important example of a non-circular duct flow. The ducts in a plate heat-exchanger are sometimes of this shape, as are the air-ducting passages in ventilating or building-heating installations. Although, in the great majority of engineering examples, the flow is turbulent, laminar flow does occur in a number of practical cases, particularly in connection with the cooling of small-scale pieces of equipment or in lubricant-flow passages.

When the fluid is Newtonian and the viscosity is uniform, an analytical solution of the momentum equation (4.1) may be obtained in terms of an infinite series of products of functions of x^* ($\equiv x/w_{duct}$) and y^* ($\equiv y/h_{duct}$). The solution, first obtained by Boussinesq in 1868 (ref. [4.2]) and quoted in many modern textbooks, including ref. [4.3], may be expressed in the form;

$$W = \frac{h^2}{\mu}\frac{dp}{dz}\left\{\frac{1}{2} - \frac{y^{*2}}{2} - \frac{16}{\pi^3}\sum_{n=0}^{\infty}\frac{(-1)^n}{(2n+1)^3}\frac{\cosh(2n+1)\pi A(x^*-1)/2}{\cosh(2n+1)\pi A/2}\cos(2n+1)\pi y^*/2\right\} \quad (4.2)$$

where A ($\equiv w_{duct}/h_{duct}$) is the *aspect ratio*. Integration of this expression allows the average velocity $\overline{W}$ to be calculated, and the friction factor to be deduced, using the formula

$$f \equiv \frac{\tau_{wall}}{\frac{1}{2}\rho\overline{W}^2} \equiv \frac{A\ dp/dz}{\frac{1}{2}\rho P\ \overline{W}^2} \quad (4.3)$$

where P is the perimeter of the duct.

The existence of the exact solutions enables the accuracy of the TEACH–C predictions to be checked.

3.1.2 *Objectives*

To explore the behaviour of fully-developed flow in rectangular-sectioned ducts in order to

(*i*) determine, for a fixed cross-sectional area and pressure gradient, the effect of aspect ratio on the velocity distribution;

(*ii*) obtain the variation of friction factor with Reynolds number and aspect ratio, and compare this with the exact solution;

(*iii*) verify that the numerical solutions with the chosen grid are in adequate agreement with the exact analytical solutions for the velocity distribution and friction factor.

3.1.3 *Method*

In this particular situation we shall take full advantage of the available flow symmetries by confining the calculations to the quadrant of the duct cross-section illustrated in fig. 4.2, for which the lower boundary is one of zero shear stress. (The dimensions of this domain, it should be noted, are $w = w_{duct}/2$ and $h = h_{duct}/2$.) A non-uniform grid is to be used, with NI=NJ=12, FEXPX = 1.1 and FEXPY = 1.0/FEXPX.

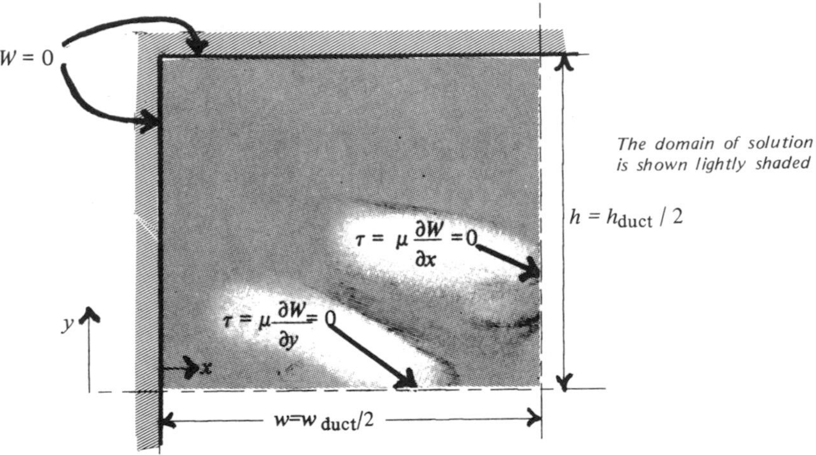

Fig. 4.2 Solution domain and boundary conditions for LESSON 1

The effect of aspect ratio at constant cross-sectional area and pressure-gradient is examined in the following way; we set the product $w_{duct} \times h_{duct}$ equal to unity, thereby fixing the area, though the precise value is immaterial; we then ascribe various values to w_{duct} and calculate h_{duct} as its reciprocal. The aspect ratio is then simply $A = w_{duct}/h_{duct} = w_{duct}^2$. Thus we are able to specify the aspect ratio simply by setting the width of the duct. First run the program as it stands, with $w_{duct} = 1.0$; then make additional RUNs with w_{duct} equal to $\sqrt{0.5}$, $\sqrt{2}$, 2.0 and 3.0.

The influence of parameters other than aspect ratio is determined by reverting to a duct-width of unity, and making RUNs which involve the variation of just one of the following quantities to the value(s) specified;

(*a*) $dp/dz = -200$ and -1000
(*b*) $\mu = 0.1$
(*c*) $\rho = 0.1$.

3.1.4 *Programming*

The major changes required from the Base Case specification are confined to the insertion of instructions for calculating the "exact" velocity distribution and friction factors from eqns. (4.2) and (4.3) respectively and modifying the boundary conditions to incorporate the additional symmetry condition. Table 4.3 shows the new FORTRAN variables needed for this LESSON. Table 4.4 summarises the changes required. This table is followed by a listing of the UPDATE needed to implement the changes.

Table 4.3 New FORTRAN variables for PROBLEM 4 LESSON 1

Variable	Quantity	Significance
EXFACT		Friction factor calculated from exact solution
EXWAVE		Average velocity calculated from exact solution

Table 4.4 Programming instructions for LESSON 1

Subroutine	Chapter	Change
CONTRO	0	♦ Provide heading for exact solution
	1	♦ Set NI = 12 and HDUCT = 1./WDUCT ♦ Set expansion factors and SORMAX
	3	♦ Set boundary conditions on symmetry plane
	4	♦ Delete south wall shear-stress calculation ♦ Calculate and print exact solutions for velocities and friction factor ♦ Amend Formats
PROMOD	1	♦ Introduce symmetry boundary at south wall

PROBLEM 4 LESSON 1 UPDATE for effecting the changes of table 4.4

```
*IDENT,P4L1
*D,CONTRO.43
      1      /'EXACT ','SOLUTI','ON    ',3*'      '/
*D,PROBLEM4.3
      NI=12
*D,CONTRO.58
      NJ=12
*D,PROBLEM4.4
      W=1.0
*D,CONTRO.60,62
      H=1./W
      FEXPX=1.1
      FEXPY=0.9
*D,CONTRO.84,85
      IMON=6
      JMON=6
*D,PROBLEM4.24,25
      FLOW=4.*FLOW
      AREA=4.*AREA
*D,PROBLEM4.27
      DH=AREA/(H+W)
*D,PROBLEM4.42
C-----THE SOUTH @WALL@ SHEAR IS ZERO IN SYMMETRY CASE
      SSS(I)=0.
*D,PROBLEM4.45
      SSAVE=-FORCET/4./(H+W)
C-----CALCULATION OF VELOCITY DISTRIBUTION FROM EXACT SOLUTION.
      DO 403 J=2,NJM1
      DO 404 I=2,NIM1
      A=0.5*H*H-0.5*Y(J)*Y(J)
      DO 405 K=1,20
      C=EXP((2.*(K-1)+1.)*PI*0.5*(X(I)-W)/H)
      CD=EXP((2.*(K-1)+1.)*PI*0.5*W/H)
```

```
      B=-16.*H*H/(PI**3)*(-1)**(K-1)/(2.*(K-1)+1.)**3*(C+(1./C))/(CD+(1.
     1/CD))*COS((2.*(K-1)+1.)*PI*0.5*Y(J)/H)
      A=B+A
  405 CONTINUE
  404 T(I,J)=-DPDZ*A/VISCOS
  403 CONTINUE
      CALL PRINT(1,1,NI,NJ,IT,JT,X,Y,T,HEDS)
*I,PROBLEM4.49
     1//27X,54HLESSON 1..LAMINAR FLOW IN RECTANGULAR SECTIONED DUCTS.//
*D,PROMOD.30,32
```

An UPDATE of the following kind will be required for the RUNs of LESSON 1:

```
*IDENT,P4L1R1
*D,PROBLEM4.4
      WDUCT=SQRT(2.)
*D,PROBLEM4.8
      VISCOS=Ø.1
*D,PROBLEM4.1Ø
      DENSIT=Ø.1
*D,PROBLEM4.15
      DPDZ=-1ØØØ.Ø
```

3.1.5 *Analysis of results*

(*i*) Plot on graph 4.1 for an aspect ratio of unity the variation of the average velocity with the viscosity, the density of the fluid and the streamwise pressure gradient. What power-law dependence of $\overline{W}$ on each of these parameters is indicated? Is the result in line with equation (4.2)?

(*ii*) Show on graph (4.2) the variation of friction factor with Reynolds number for all the RUNs made with an aspect ratio of unity. Do the predictions support the conclusion that for a given aspect ratio the friction factor is a function only of the Reynolds number? What power-law dependence is indicated?

(*iii*) Also plot on graph 4.2 the friction factors for the RUNs made with aspect-ratios different from unity. Does the use of "hydraulic diameter" as a scaling length bring the friction factors on to a single line? What conclusions do you draw?

(*iv*) Plot on graph 4.3 the variation of flow-rate with aspect ratio. What conclusions may be drawn about the effect of aspect ratio on flow-resistance? Explain.

(*v*) Plot on graph 4.4 the variation of $W/\overline{W}$ along the mid-plane of the duct for the different aspect ratios examined. Do the predictions for A = 2.0 and 0.5 display the expected mirror-symmetry? Above what aspect ratio does the velocity variation across the narrower gap agree closely with the following analytical result for the flow between parallel planes:

$$W/\overline{W} = \frac{3}{2}(1 - \frac{y}{h})^2$$

(*vi*) What are typically the differences between the predicted velocities and friction factors and those given by the exact solution?

3.2 LESSON 2 Laminar flow in ducts of non-circular cross-section

3.2.1 *Physical background*

It often happens in practical circumstances that flow passages are, by accident or design, neither circular nor rectangular in cross-section. Fig. 4.3 below illustrates one such situation, where the matrices of a compact heat-exchanger have passages of approximately triangular section. Typically, the designer of this equipment needs to know how the heat transfer rate and pressure-drop vary with the flow rate, the shape of the duct and the fluid properties, so as to enable him to optimise the performance.

Provided that conditions are such that fully-developed laminar flow prevails, analytical methods are available for solving eqn. (4.1) for such passages. However, these methods are often complex and even for simple shapes the results are usually only expressible in the form of infinite series (cf. eqn. (4.2), which gives the solution for a rectangular duct), so that numerical methods are often preferable.

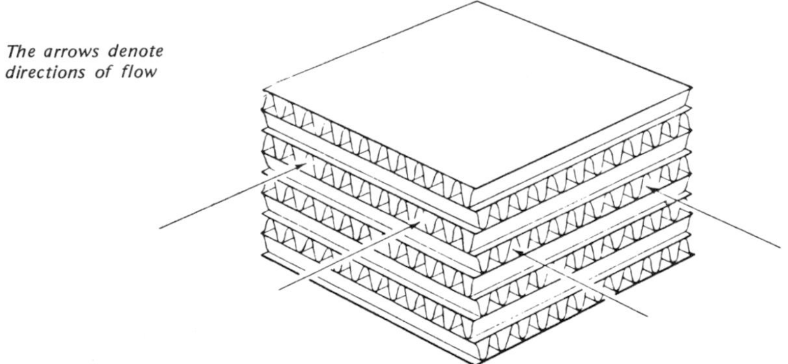

Fig. 4.3 Illustration of the flow passages of a typical compact heat-exchanger

The present LESSON is concerned with obtaining solutions for the duct configuration of fig. 4.4; here, intrusions of dimensions $x_B \times y_B$ have been introduced into two corners of an otherwise rectangular duct. This shape has been chosen as being particularly simple to incorporate

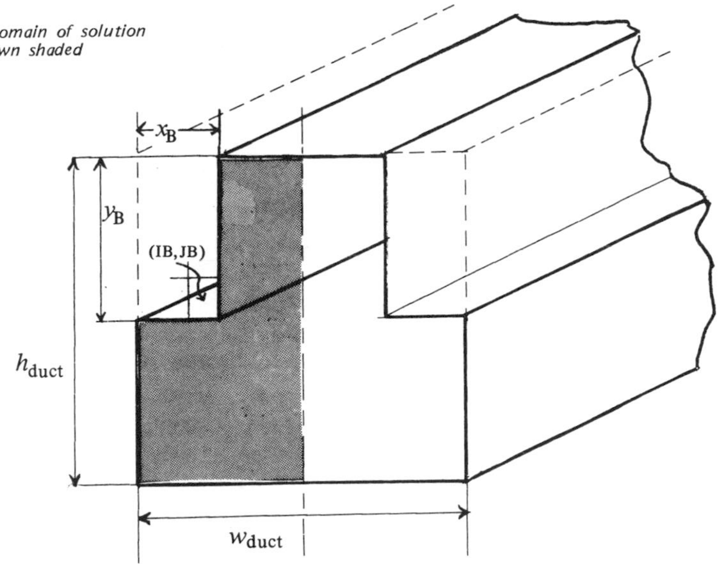

Fig. 4.4 Illustration of shape of flow passage considered in LESSON 2

into the TEACH–C program.

3.2.2 *Objectives*

To determine, for the duct of fig. 4.4, the dependence of :

(*a*) flow rate on the size of the intrusion, for a prescribed flow area and pressure gradient

(*b*) friction factor on Reynolds number, for a fixed size of intrusion.

3.2.3 *Method*

Calculations are to be made for a duct of equal external dimensions (i.e. $w_{duct} = h_{duct}$) and fixed cross-sectional area, the latter being arbitrarily specified as unity. A uniformly-spaced grid is employed, with with NI = NJ = 12; the extent of the intrusion is specified through the indices IB and JB which denote the exterior cell at the junction of the intrusion walls (see fig. 4.4). Once the proportions of the intrusion are fixed, it is a straightforward matter to work out the external dimensions which yield the required cross-sectional area :

TEACH–C PROBLEM 4 LESSON 2

instructions for doing this are incorporated into the program modifications described below. The standard TEACH–C procedures are used to insert the appropriate wall-shear-stress * formulae at interior cells adjoining the intrusion and to set the velocities at exterior locations to zero.

The computations required for objective (*a*) consist simply of making predictions for various sets of values of IB and JB so as to yield different sizes of intrusion. Table 4.5 below contains some recommended settings which give rectangular intrusions occupying proportions of the nominal cross-section (defined as $w_{duct} \times h_{duct}$) ranging from zero to 0.3. The 'zero-intrusion' case is included as a basis for comparison.

IB	JB	Ratio of area of intrusion to nominal cross-section
1	12	0:1
3	8	8:100
4	6	18:100
5	4	32:100

Table 4.5 Recommended sets of values of IB and JB

For the exploration of friction factor (objective (*b*)) two additional computations are required, in which the size of intrusion is fixed at the largest value specified in the above table, and the pressure gradient dp/dz is changed to 0.1 of, and then 10.0 times, the original value.

3.2.4 *Programming*

The main alterations to the Base Case consist of providing, in CONTRO, sequences for calculating w_{duct} and h_{duct} and amending the sequences there for calculating geometry-dependent quantities such as hydraulic diameter, flow rate, shear-stress, etc. Changes are also required in PROMOD to incorporate the effects of the intrusion. These and other alterations are summarised in table

* see the TEACH–C Guide

4.7 below, which is followed by a listing of the UPDATE which effects them. The new FORTRAN variables introduced are listed and defined in table 4.6.

Table 4.6 New FORTRAN variables for LESSON 2

Variable	Quantity	Significance
AINT		Area of intrusion
IB		I index of cell in corner of intrusion
JB		J index of cell in corner of intrusion
XB	x_B	Dimension of intrusion
YB	y_B	Dimension of intrusion

Table 4.7 Programming changes for LESSON 2

Subroutine	Chapter	Change
CONTRO	0	♦ Put IB and JB into COMMON
	1	♦ Set NI=12
		♦ Set IB and compute JB and duct dimensions such that area of duct is unity
		♦ Reset vertical traverse limits around intrusion
	2	♦ Calculate dimensions of intrusion and source-normalisation factor
		♦ Calculate area of intrusion
	4	♦ Modify calculation of area of duct and of flow rate
		♦ Modify shear-stress calculations at boundaries of intrusion
		♦ Modify Formats
PROMOD	0	♦ COMMON block as in CONTRO
	1	♦ Delete momentum source term within intrusion
		♦ Set correct north boundary condition at intrusion
		♦ Insert west coundary condition at intrusion

LESSON 2 UPDATE for effecting the changes in table 4.7

```
*IDENT,P4L2
*I,PROBLEM4.1
      1/P4L2/IB,JB
*D,PROBLEM4.3
       NI=12
*D,PROBLEM4.4,CONTRO.60
C------SPECIFY INTRUSION S.T. OPEN AREA OF DUCT HELD CONSTANT AT UNITY.
       IB=3
       JB=NJ-2*(IB-1)
       H1=FLOAT((IB-1)**2)
       H2=FLOAT(NI-2)*FLOAT(NJ-2)
       H=SQRT(1./(1.-2.*H1/H2))
       W=0.5*H
*I,CONTRO.82
       IF(I.LE.IB)JN(I)=JB-1
*D,PROBLEM4.13
       XB=XU(IB+1)
       YB=H-YV(JB)
       SNORM=-DPDZ*(W*H-XB*YB)
*I,PROBLEM4.14
       WRITE(6,2190)XB,YB
*I,CONTRO.178
C-----IMPOSE SYMMETRY CONDITION
       DO 403 J=2,NJM1
  403 T(NI,J)=T(NIM1,J)
*I,PROBLEM4.19
       IF(I.LE.IB.AND.J.GE.JB)GO TO 411
*I,PROBLEM4.22
  411 CONTINUE
*I,PROBLEM4.33
       IF(J.GE.JB)SSW(J)=-VISCOS*(T(IB+1,J)-TLEFT)/(X(IB+1)-XU(IB+1))
*I,PROBLEM4.40
       IF(I.LE.IB)SSN(I)=-VISCOS*(T(I,JB-1)-TTOP)/(YV(JB)-Y(JB-1))
*I,PROBLEM4.49
      1 ,54X,'LESSON  2'//,38X,'RECTANGULAR INTRUSION IN THE DUCT '//
*I,CONTRO.194
 2190 FORMAT(1H0,' INTRUSION OF DIMENSIONS XB=',1PE12.3,2X,'YB='
      1,1PE12.3)
*D,PROBLEM4.58
       DO 1310 J=JS(IL),JN(IL)
*I,PROMOD.17
       NJM1=JN(IL)
*I,PROMOD.18
C-----BOUNDARY CONDITIONS AT INTRUSION
       RDYN=RV(JB)/(YV(JB)-Y(JB-1))
       IF(IL.GT.IB) GO TO 202
       J=JB-1
       AN(J)=0.0
       DN=GAMH(IL,J)*SEW(IL)*RDYN
       SU(J)=SU(J)+DN*T(IL,JB)
       SP(J)=SP(J)-DN
  202 CONTINUE
       DO 203 J=JB,NJM1
       DXW=X(IB+1)-XU(IB+1)
       IF(IL.NE.(IB+1))GO TO 204
       AW(J)=0.0
       DW=GAMH(IL,J)*SNS(J)*RY(J)/DXW
       SU(J)=SU(J)+DW*T(IB,J)
       SP(J)=SP(J)-DW
  203 CONTINUE
  204 CONTINUE
*I,PROBLEM4.56
      1/P4L2/IB,JB
```

An UPDATE of the following kind will be required to perform the RUNs of LESSON 2

```
*IDENT,P4L2R1
*D,P4L2.4
      IB=5
*D,PROBLEM4.15
      DPDZ=-1ØØØ.Ø
```

3.2.5 *Analysis of results*

(i) On graph 4.5 plot the variation of the ratio of the flow rate with an intrusion to that without, as a function of the ratio of areas $x_B \times y_B / (w_{duct} \times h_{duct})$. Explain the trend you observe (*hint*: plot also the ratio of the perimeter of the duct with intrusion to that without, and deduce the implications with regard to wall shear force). What practical conclusions may be drawn about, say, the design of a lubricant passage for a bearing?

(ii) Sketch on the computer outputs and then on graphs (4.6) and (4.7) contours of constant velocity (*isovels*) for the cases of no intrusion and the maximum intrusion, both for the same value of dp/dz. What is the effect on the flow distribution of introducing the intrusion?

(iii) Plot on graph 4.8 the predicted variation of friction factor with Reynolds number based on hydraulic diameter for the explorations of objective (*b*). Is a power law dependence indicated? If so, what is the power? Plot also on the same graph the friction factor predictions of objective (*a*), using different symbols for each case. What conclusions may be drawn regarding the validity of the hydraulic-diameter concept for these flows?

Graphs for LESSON 1

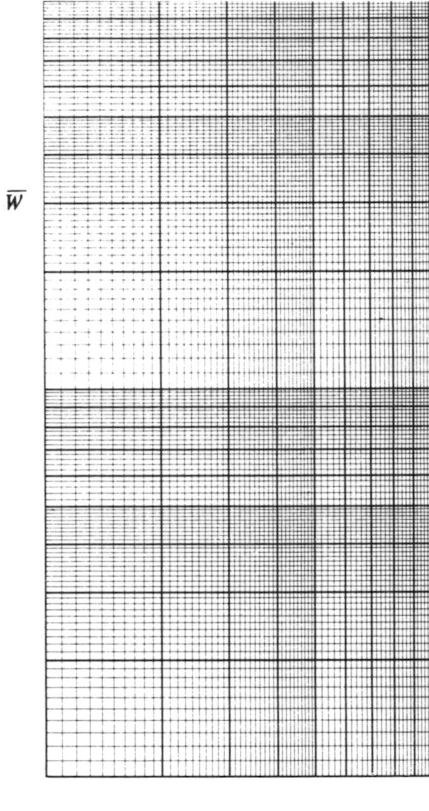

Graph 4.1 Average velocity plotted against viscosity, density and pressure gradient

(LESSON 1)

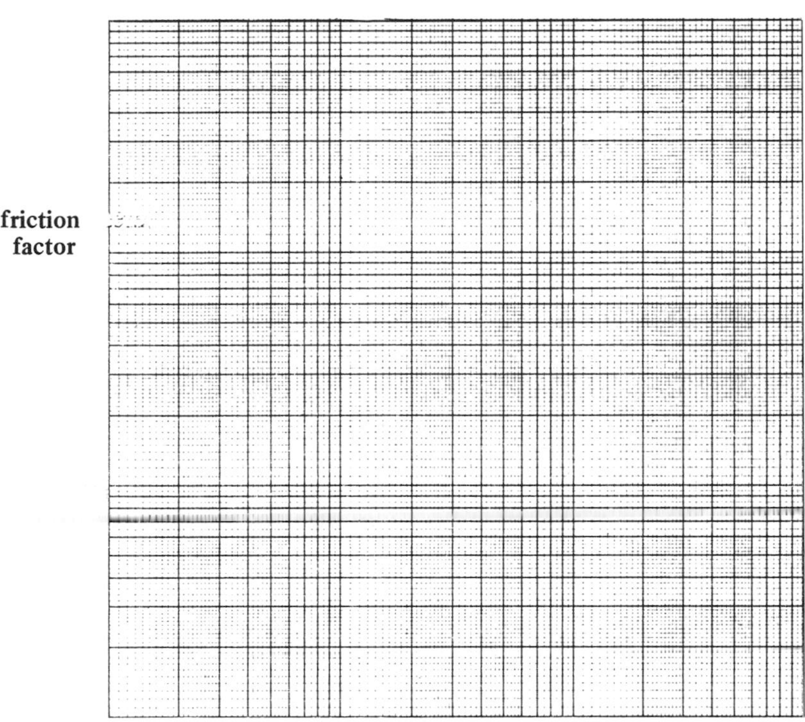

Graph 4.2 Variation of friction factor with Reynolds number

(LESSON 1)

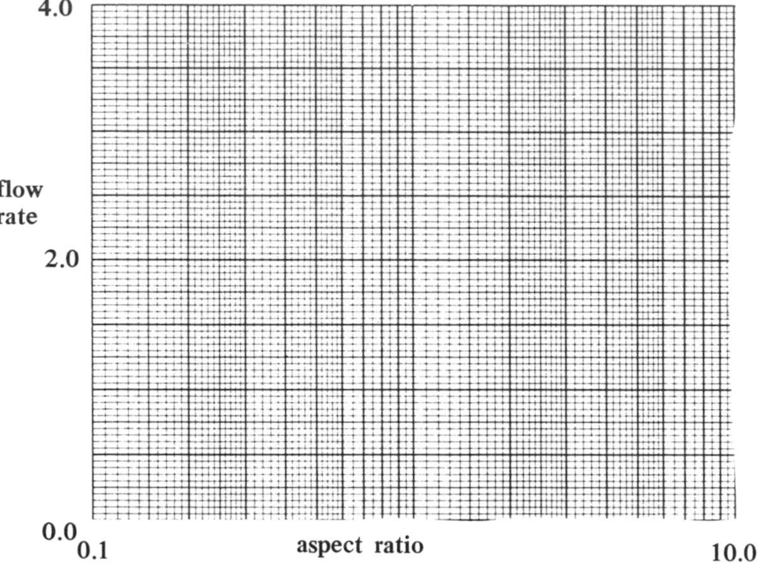

Graph 4.3 Variation of flow rate with aspect ratio (LESSON 1)

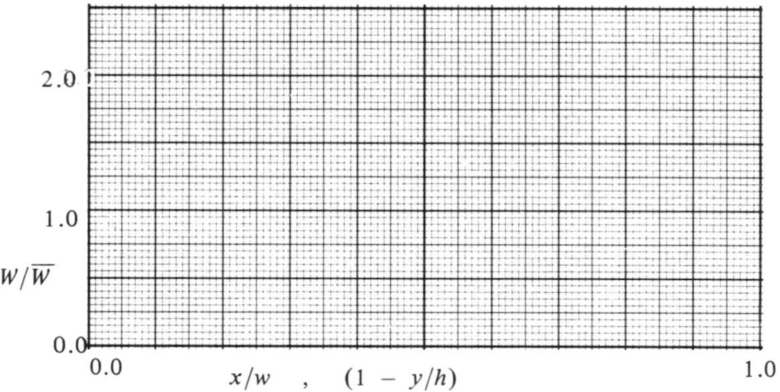

Graph 4.4 Variation of velocity along the mid-plane of the duct for various aspect ratios (LESSON 1)

Graphs for LESSON 2

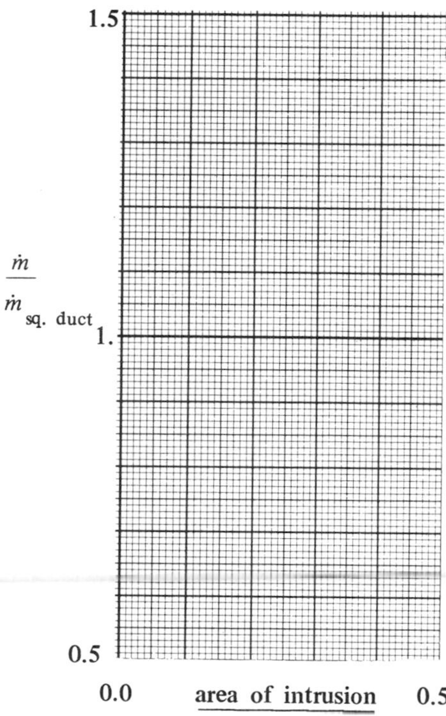

Graph 4.5 Variation of normalised flow rate with ratio of areas

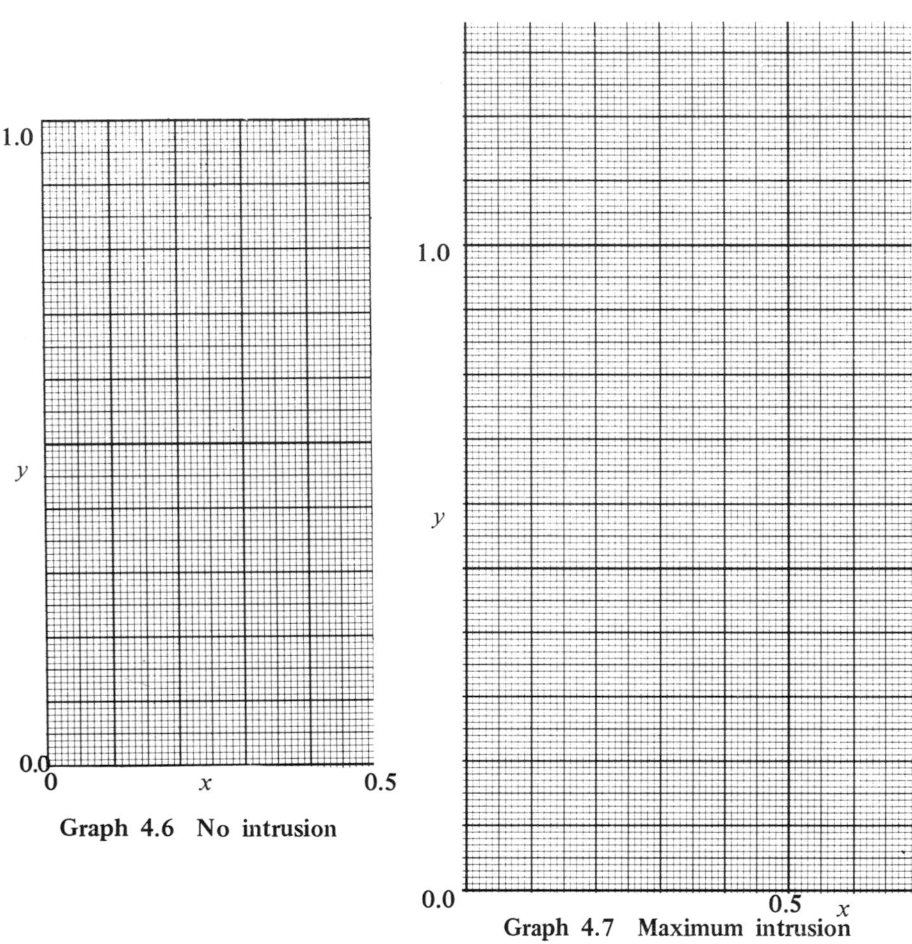

Graph 4.6 No intrusion

Graph 4.7 Maximum intrusion

Contours of constant velocity (LESSON 2)

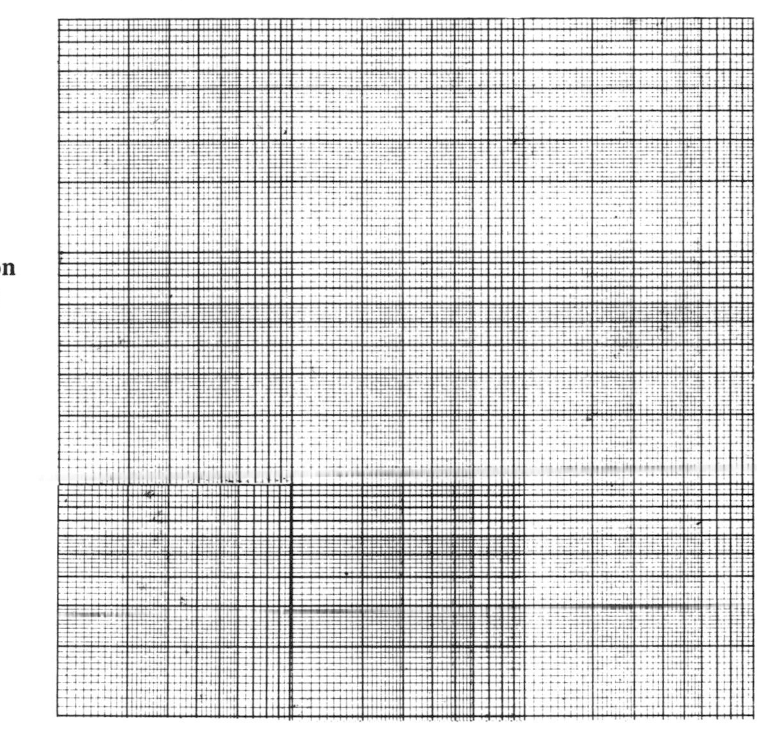

Graph 4.8 Variation of friction factor with Reynolds number (LESSON 2)